Mountains to Metropolis:

The Elbow River Watershed

Diane Coleman

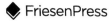 FriesenPress

Suite 300 - 990 Fort St
Victoria, BC, Canada, V8V 3K2
www.friesenpress.com

Copyright © 2015 by Diane Coleman

First Edition — 2015

All rights reserved.

No part of this publication may be reproduced in any form, or by any means, electronic or mechanical, including photocopying, recording, or any information browsing, storage, or retrieval system, without permission in writing from FriesenPress.

ISBN
978-1-4602-7115-5 (Paperback)
978-1-4602-7116-2 (eBook)

1. Nature, Environmental Conservation & Protection

Distributed to the trade by The Ingram Book Company

www.elbowriverwatershed.com

Table of Contents

Preface — vii
Acknowledgements — ix
Introduction — xi

Part 1. In the High Mountains

1. Starting at the Source — 3
2. Moving Down the Mountain Stream — 15
3. Where Mountains Meet Foothills — 31

Part 2. The Foothills Between

4. Traversing the Great Foothills Ridges — 51
5. Meeting the Family — 71
6. Recreating at the River — 91
7. Historic Hamlet on the Elbow — 111

Part 3. To Plains and Metropolis

8. Horses, Hunters and Homesteaders — 133
9. Warriors in the Watershed — 157

10. Missions, Cowboys and an Urban Thirst 171

Part 4. Whither the Elbow?
   11. Reflections 195

Appendices
   1. Historical Timeline:
      Elbow River Watershed and Region 209
   2. Scientific Names 217

References 221
Glossary 239
Index 247

A watershed is a living, pulsating being. Put your ear to it, and hear it hum with life. Put its watershed sisters on either side of it, join them with their larger mothers, and soon you have our living, pulsating Earth – nothing less.

Every watershed has a story. This is the Elbow's.

Preface

This story began for me when I moved into the Elbow watershed in the mid-70s, two blocks from the river in a crowded neighbourhood close to Calgary's downtown. The river provided a place of refuge, a forested "blue" space which seemed far from the noise and bustle of the streets. There we could walk the dog, paddle or swim in the water, climb the trees and skate or cross-country ski in winter. Then, the river was not part of its whole — its watershed — but rather a beautiful place near which we were fortunate to live. But that beautiful place is truly so much more.

Water has always been an attractant for me. As a child, any stream, creek or pond grabbed my interest and drew me to it — to splash, to wade, to swim or just to sit on the edge and enjoy the ripples and the water creatures coming and going. The broader landscape was also fascinating and I went on to work with it as a geographer. It was natural for me, after living by the Elbow River for three decades, to belatedly begin to focus on the river and on its broader watershed. When I moved from the city, where potable water magically comes out of the tap and wastewater disappears with a flush, to the middle watershed, our water was provided by a small water co-operative and we had our own septic system to maintain. This afforded a different reality — an opportunity to think about where our water comes from, how it is used and where it goes when we have finished with it.

What is a watershed really? What are its parts, how does it function, is it so important to keep it healthy, and how do we do that? The Elbow River watershed is under pressure. It appears diminutive, yet as the city grows and expands, this proximate watershed is experiencing ever-increasing demands for industrial, urban and recreational development. I now know that the integrated functioning of any watershed is based on the health of all of its parts; I hope to help ensure the Elbow's continued health through understanding and thence good management.

Consider this more of an amble through the Elbow watershed than a scientific treatise. There are many other books which discuss the Elbow River and its watershed in various contexts, including hiking guides, descriptions of the building of the Glenmore Dam, or the history of First Nations in and around the watershed. Many of these are included in the References, and I encourage you to seek them out for more detailed information. Also, when reading this book, be sure to consult the historical timeline in the Appendices, as it provides a useful context for the events described in the text. I hope you enjoy it all.

Acknowledgements

I owe much of this story to my hiking friends — SanDee, Heather, Marjorie, Gail, Deb, Anne and Sheilagh — who were willingly dragged into the Elbow watershed on so many occasions. Also to Guy who headed me down the right format route, to the Elbow River Watershed Partnership who loaned me materials, to Donna and Tracy for their horse insights, to Bob, Heather and Guy for their photographs, and to the friends and family who kindly inquired about my progress along the seemingly endless journey. Dr. Lynn and Arne carefully read the manuscript. Joan Dixon ably whipped the text into shape. Robin and Dexter advised on the maps, Soo Kim did the diagrams and the Bow River Basin Council allowed the use of their Elbow watershed map. My husband Bryan provided the most — moral support, a hug during the grumpy periods, and the ability to often stop talking when I needed to think. Thanks, everybody!

Introduction

At its cold glacier source in the high eastern ranges of the Rocky Mountains, the Elbow River is just a small alpine stream. It tumbles down through the rocky terrain above tree line, then widens as it follows its winding valley through thick boreal forests to the aspen parklands and grazing lands. It meanders past farms and acreages, until it flows through a placid reservoir into the crowded city, and finally to its confluence with the larger Bow River among office towers and condominiums. Its modest watershed covers about 1,200 square kilometres, one-fifth of the size of Banff National Park. The river itself is only 120 kilometres long, shorter than the drive from Calgary to Banff. It not only provides drinking water for one in seven Albertans, but also critical wildlife habitat and significant residential, recreational, agricultural and industrial opportunities. This watershed assumes an importance that belies its size.

Some Words about Watersheds

A watershed, sometimes called a drainage basin, is a land area that funnels all the water falling onto, and draining from, the land to a common outlet — such as another waterway or a water body. The Elbow watershed thus includes all the area drained by tributaries that contribute water to the Elbow River, and its outlet is the larger Bow River. A watershed is separated from its neighbours by topographic divides like ridges, hills or mountains. Within a watershed, the water may take the form of wetlands, rivers and streams, lakes and ponds, reservoirs and groundwater.

Before the water reaches the watershed's outlet, however, it is subject to many influences: it may be absorbed by vegetation, consumed by people, walked in by animals, and evaporated by the sun. What happens within this watershed, then, affects the quality and quantity of the water in it and leaving it — the watershed exerts an integrating force.

Why should we care? Because each of us lives in a watershed. It is to our benefit that we understand the interdependent factors that determine the quality and quantity of the water that sustains us there. Each watershed is different — the integrated result of its physical elements, its history, its land management — and the resulting water flow and water quality within the watershed are a result of these complex conditions. In other words, all of those conditions upstream affect conditions downstream. Sustainable watershed management should be a concern of everyone within and downstream of their watershed.

The term watershed can also be used in a figurative sense, implying a crucial dividing point in time. This may have been what the respected Alberta historian, author and statesman Grant MacEwan meant in naming his last book, *Watershed: Reflections on Water*. As he stated,

> The care and responsibility we show for our water demonstrates much about our values, including our level of concern for the quality of life of future generations.

The figurative meaning is relevant for the Elbow watershed. This unassuming watershed has many managers for its water and other resources, including the Kananaskis Improvement District, Alberta Parks, Rocky View County, the Tsuu T'ina Nation and the City of Calgary. A host of other stakeholders are also involved, from local farmers and ranchers to forestry companies to the hamlet of Bragg Creek. Such management complexity has led, as in some other watersheds, to the formation of a watershed partnership to more effectively deal with its issues in a collaborative manner — a promising approach.

The Elbow River watershed is at that critical "watershed" point in time. The City of Calgary is experiencing unprecedented economic growth. Residential expansion into the agricultural and natural areas of the watershed is increasing rapidly, with a projected population

Introduction

growth of more than 30,000 in the next decade. Water quality in the Elbow River and its aquifer upstream from Calgary has shown a significant decrease — a real concern for those who depend on this water source. Accelerated expansion of logging in the watershed is planned over the next 20 years. Ever-increasing tourism in the region puts much greater pressure on the already popular recreation facilities within the watershed, most just an easy hour's drive from the growing city. Added to all this, the epic flood of June 2013 not only physically changed much of the Elbow watershed, but led to significant policy and program changes affecting the watershed and its residents, human and non-human alike.

In this book, I explore the history of this beautiful watershed, its physical nature, and its cultural and ecological components — at a time when more knowledge about this organism can be instrumental in planning and understanding our relationship with it. My own experiences in the watershed — from hiking up to its source mountain glacier, to daily life beside the river, to seeking solace in the green space near its urban endpoint — provide the background against which other watershed stories are presented. Grizzly bears and mule deer, park wardens and cowboys, First Nations and first settlers, range cattle and coyotes, urbanites and beavers, city engineers and soldiers, Grey Nuns and missionaries — all are part of this watershed. And their stories reflect how they have shaped and been shaped by the physical and spiritual power of the river at the watershed's core. These are all part of the Elbow watershed's ongoing story.

Part 1.

In the High Mountains

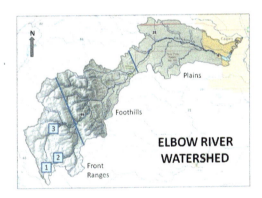

(map courtesy Bow River Basin Council)

Chapter 1.

Starting at the Source

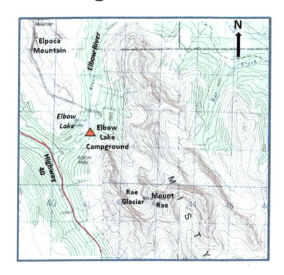

Map showing the source of the Elbow River at Rae Glacier and Elbow Lake

Just like Alice in a different Wonderland, I was told that when you want to understand something, begin at the beginning. Good advice for life, and possibly for watersheds too? For years my hiking group had been exploring the watershed of the Elbow River, but had yet to visit the source of this beautiful watercourse. Today, on a bright blue mountain morning, we are setting out on a backpacking jaunt and something of a quest. We hoist up the deadweight backpacks, grab bear spray and climb up the steep gravel track until we reach the high point of the trail — Elbow Pass. Finally, in front of us, is the welcome sight of little Elbow Lake glistening in the sunshine, and of the quiet backcountry campsite on its southeast margin.

Elbow Lake, at Elbow Pass

Here at the westernmost point of the Elbow watershed, Elbow Lake is fed by springs and the creek that comes from the highest source — Rae Glacier — another 600 metres up on the northern slopes of Mount Rae. At 3,218 m, the mammoth, block-shaped Mount Rae, with its tightly-folded limestone rock layers, is one of the highest mountains in the Front Ranges of the Rocky Mountains, and the highest in the Elbow watershed. Its notched peak can even be seen from the prairies, 60 kilometres to the east. Mount Rae is the northernmost mountain in the spectacular 17-km-long Misty Range. This range was named in 1884 by George Mercer Dawson (then assistant director of the Geological Survey of Canada) following an extended period of bad weather in the area.

Starting at the Source

Rae Glacier, on Mount Rae (Photo courtesy of Robert Lee)

To reach our destination, we walk single-file on a rooty path along the quiet southeast shore of the lake, and turn east to begin the three-kilometre climb up to the glacier. The trail winds through a thick subalpine forest of Engelmann spruce and lodgepole pine, and then emerges into open rocky terrain with a clear view of the mountain and glacier ahead. Beside the trail runs a small, cold mountain stream — the nascent Elbow.

We turn and look back down the valley toward Elbow Lake, enjoying the extended vista of the "rip-saw" mountains of the Opal Range stretching out into the northern distance. The Opal Range was also named by George Dawson, in this case for quartz-lined rock cavities that he thought were opal, but

Mount Rae and Dr. John

Mount Rae was named in 1859 by the celebrated explorer of western Canada, Captain John Palliser, for the less-known Arctic explorer, Dr. John Rae. Scottish by birth, he worked for the Hudson's Bay Company and was hired to survey nearly 2,500 kilometres of the Arctic coast. He learned Arctic survival from the Inuit, covering over 10,000 kilometres on snowshoes. He unearthed the tragic 1847 fate of Sir John Franklin, the news of which had mixed reactions in England. (It was later proved that Rae, not Franklin, found the final link to the Northwest Passage.) In 1864 Rae carried out his last major expedition — to survey a route for a telegraph line through the Canadian Rockies. Magnificent Mount Rae is a worthy namesake for this consummate surveyor and traveller.

which disappointingly turned out to be silica embedded with quartz. Its nearly vertical beds of Rundle limestone have created spectacular mountain scenery, and significant challenges for mountain-climbers; many of these peaks were not scaled until the 1950s. I can see why: they look daunting to me.

View looking north from Rae Glacier down to the main Elbow valley

Below us, the impressive Elpoca and Tombstone mountains can be seen looming over the forested valley. Mounts Rae and Elpoca surround the Elbow's source, forming part of the divide between the Elbow and the Kananaskis watershed to the west. Elpoca, the southernmost mountain in the Opal Range, stands slightly lower than Rae, at 3,029 m. Named for its position on the divide between the **El**bow and **Poca**terra watersheds (get it?), its vertically-dipping limestone strata give Elpoca its distinct thrusting appearance.

Extending northward from Elpoca, seven other mountains of the Opal Range — Jerram, Burney, Blane, Brock, Head, Packenham and Evan-Thomas — guard the western edge of the Elbow watershed. Six of these giants were officially named in 1922 to honour admirals of the British Royal Navy who had participated in World War I. The remaining mountain — Mount Head — was given its name by Palliser in 1859, to recognize Sir Edmund Walker Head, then the governor general of the province of Canada.

As we stand admiring the mountain panorama, we wonder, like so many others when encountering the towering Rocky Mountains, about the origins of these massive structures. As with landscapes worldwide, geological formations underlie and are largely responsible for the present physical structure of this area. Over time, three consecutive phases of landscape construction have taken place: first, deposition of the sediments and

Whence Came the Rockies

Over a couple of billion years in the Earth's early history, sediments from the North American continent were deposited on its ocean shelf to the west, settling in near-horizontal layers and gradually compressing under their own accumulated weight into sedimentary rock. Then, a mere 170 million years ago, something began to move in the earth's crust. Under great pressure, massive thrust sheets of rock shoved eastward onto the North American plate, one and then another moving on top of each other until the whole west side of the continent was a mass of uplifted jumbled rock. These were the earliest Rocky Mountain ranges.

West-east cross-section through the Rocky Mountains, showing the results of mountain-building

Where our Interior Plains are now, these movements forced the existing rock layers to bend downwards, forming a giant basin that soon filled with ocean water — an inland sea. Over time, eroded sediments from the surrounding lands accumulated in this basin until it became a swamp, filled with dinosaur life. Meanwhile, crustal movements to the west continued. The craggy Main Ranges of the Rockies were lifted even higher, then the Front Ranges further east, and in a final push, the Foothills east of them. Finally, only 50 million years ago, mountain building ceased, and erosional forces of wind, water and ice began to tear down what had been built. The Rocky Mountains we see today are the result.

The Power of Ice

In the creation of this modern mountain and foothills landscape, the third, erosional stage began with the most recent extensive ice age in North America during the Pleistocene epoch (beginning 1.6 million years ago and ending 10,000 years ago). Two opposing forces drove toward the area of the future Elbow watershed. Centred near the present Hudson Bay, the Laurentide Ice Sheet pushed west to the Calgary area. Meanwhile, accumulating snow over the Rocky Mountains, compacted into ice as the massive Cordilleran Ice Sheet. At its eastern extent, it sideswiped the Laurentide Ice Sheet and veered to flow southwards. Part of that meeting occurred right within this little Elbow watershed, just west of Calgary. ➔

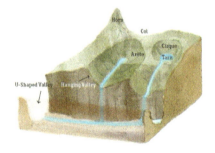

Features formed by alpine glaciers

their consolidation into rock; next, uplift of the rocks; and finally, erosion by water and glacial ice. The basic mechanism for mountain building in this area is orogeny, in which rigid interlocking slabs of the Earth's crust (or plates), floating on a semi-fluid interior layer, collide with, or slide under, each other. Such interaction can lead to such violent events as earthquakes, volcanic eruptions near the plate boundary, and, in the case of the Rockies, mountain-building.

The mountain-building results lie within our view, and our heads swivel almost as one toward the object of our quest — Rae Glacier. In addition to its visible contribution to the flow of the Elbow River, modest-appearing Rae Glacier has had an important role in altering the local mountain landscape. Glacial forces seem somehow more understandable than the whole mountain-building process. Maybe that's because the effects of mountain glaciation can be easily visualized in the current Rockies and other landscapes.

Diminutive Rae Glacier sits in a small cirque on Mount Rae, a remnant of a much larger resident glacier. Located in the alpine zone about 500 m below Mount Rae's summit, it has been consistently shrinking. The glacier does not look like a great subject for our cameras. At this time in July, it is not

Starting at the Source

pristine white, but dirtied by the rockfalls, dust and debris on its surface. Its tarn is a small muddy pond — not like the clear aquamarine tarns we have seen on other alpine hikes. Nonetheless, it has both an historical and a current appeal of its own.

The glacial features within the Elbow watershed are nearly all ancient, created by long-gone ice. About 130 years ago, this glacier extended half a kilometre further down the mountainside, and studies show that it has been retreating an average of six metres a year since 1916. That adds up to a lot, for a small glacier like Rae. As we follow the trail up to Rae Glacier, glacial features are everywhere. We have already climbed over tree-covered moraines, and passed by rocky lateral and smaller medial moraines closer to the glacier. This little tarn is dammed by a cobbly moraine at the bottom of the cirque, formed when the glacier advanced and then receded long ago. Looking up at the surrounding mountains, strikingly sharp ridges and peaks are everywhere — not a rounded hill in sight. And when we look back down the valley towards Elbow Lake and the campsite, the vista is one of a broad valley between high mountains — it does not take much imagination to see how a massive glacier crept slowly down that valley, carving out the landscape around it.

→ Underneath any moving glacier, the land is substantially changed, as are the surfaces ahead of and beside the glacier. As alpine glaciers moved down the mountain valleys, joining together as they grew thicker, rock surfaces were gouged and scraped, and eroded materials were plucked up and deposited elsewhere by the ice and subglacial rivers. Before this glaciation, the Rockies were gently rounded and rivers flowed in V-shaped valleys. The mountain glaciers carved away valley materials, leaving broad U-shaped valleys in their wake.

Semi-circular hollows with steep headwalls (cirques) were carved by the ice into mountain slopes, later containing small post-glacial lakes (tarns). Between back-to-back cirques, sharp ridges (arêtes) or low saddles (cols) were formed. Some mountain peaks were eroded by the ice into horns or pyramids. Where glaciation deepened a main mountain valley and sliced away the previous gradual side-valley descent to the valley bottom, aptly-termed hanging valleys were left to adjoin the main valley high above its bottom, often marked with cascading waterfalls (such as Takakkaw Falls in Yoho National Park). The result of all this? Our modern Rocky Mountains, and some of the most spectacular scenery in the world.

Moss campion

Is Global Warming an Issue Here?

Obviously the climate of the watershed has changed remarkably over time, moving as it did through long periods of glaciation and glacial retreat. How do we know that? To go back, say, a million years to look at the paleoclimate, only scientific analysis of indirect evidence is available. In some locations, ice core data provides relevant information from up to 200,000 years ago. Further back, ocean sediments provide useful data on global scale paleoclimates. Six or seven ice ages have occurred in this time period, one about every 100,000 years, with warmer interglacial periods between them. However, in the last 8,000 years, climatic conditions have been relatively stable worldwide; this is probably the case in our watershed as well. →

As the sun drops behind the high west wall of the cirque, we begin the descent to our camp at the lake below. In the afternoon shadows, the air chills quickly; we don fleeces and move faster, picking our way carefully along the rocky trail beside the flowing water. Little wildlife has been seen up here today, with the exception of piping pikas darting around in the limestone rubble. Alpine wildflowers, however, are plentiful. Moss campion, alpine forgetmenot, rock jasmine, white mountain avens and alpine buttercups cling to sparse pockets of soil between the rocks, providing rich splashes of colour on the stony surfaces.

Starting at the Source

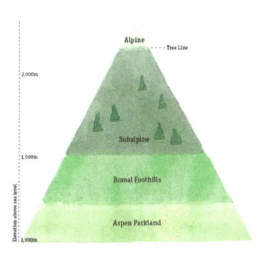

Ecozones of the Elbow River Watershed

Here in the chilly alpine zone above the treeline (about 2,100 m elevation), the mean annual temperature is low (below -4C compared to over +4C at Calgary) and the temperature range is small. This is a harsh environment for both plants and animals, and the diversity of both is as low as the temperature range. Precipitation is generally high, falling mainly as snow in winter. In the warmer summer months, the mean daily temperature may only reach 6C, usually dipping below freezing at night. Adding to the cold environment is the wind — the strong prevailing westerlies do not bring the warming chinooks to this zone, but add biting wind chill and desiccation throughout the year.

As we know to do when exploring any alpine zone — this being one of the most fragile and unique ecoregions in Alberta — we tread carefully, staying on the rocky trail. Because of the severe conditions, alpine

→ To look back only a thousand years, the writings of historians and diarists during that period, and indirect sources such as tree rings or plant pollen, can be consulted. Using such information, for example, a warm period from about 1100 to 1300 C.E. was noted, followed by a Little Ice Age between 1400 and 1800. Explorers in western Canada since the early 1800s have always noted weather-related information in their journals, more relevant data for climatic conditions in this watershed and region than European or global data. And only in the last hundred years or so have meteorological records provided accurate and consistent climate data, especially for temperature; precipitation data have been recorded much less reliably. So it is just in the modern era (since about 1860) that the historical climate and how it has changed to the present day is defined. For the Elbow watershed, the longest records in the region are those of Calgary and Banff.

The global average temperature has risen in the order of 0.5C since 1860, but this average has varied widely from decade to decade and from location to location. For example, warming during the last century was greatest in mid-latitude continental locations (including the region of the Elbow watershed) in winter and spring, and minimum daily temperatures have increased nearly twice as much as maximums. Recent studies conducted in the Calgary region have shown that the mean and minimum annual temperatures have increased about 0.9C and 1.1C respectively in the past 100 years. Forecasts for the future are continually being revised.

vegetation is low-growing, cold-hardy and sparse. Mosses, lichens, grasses and wildflowers grow sparingly on rocks and in thin soil pockets where the microclimate permits. Dwarf birch, willow, sedges and heathers can be found in wet and/or protected areas. The small wildflower, moss campion, for example, grows relatively well in the rocky soils of the alpine zone, but takes 10 years to produce its first flowers. After 25 years of growth, its circumference is only 25 cm, according to Rockies authority Ben Gadd.

Little pika, the rock rabbit (Photo courtesy of Robert Lee)

Even as the harshness of this alpine environment is recognized, it also appears to be changing. Shrinkage of the Rae Glacier is one indicator; decline of the pika population may be another. I stalk an elusive one with my camera but it seems to be enjoying teasing me, quickly popping behind rocks after piping to get our attention. Finally, I give up and he smirks from the top of a rock higher up the slope.

On the way down to our camp, I peer up at the steep rocky slopes, in hopes of seeing one of the approximately 125 resident mountain sheep perching on a narrow ridge or a golden eagle soaring above in the cloudless blue sky. Not today, however — only a hoary marmot whistles in the distance. Eventually, we re-enter the subalpine forest, cross the lumpy moraines, and reach the placid lake and the campground. Here, we can take advantage of some sun still shining on the mossy lakeshore.

Picturesque Elbow Lake is a popular destination for Calgary hikers, picnickers and fishermen, accessed after a relatively short climb up from Highway 40 in Kananaskis Country. Despite frequent grizzly bear sightings in this part of the watershed, on beautiful summer days adventurous visitors paddle or fish in its icy water or head downvalley on mountain bikes. Those wanting a relaxed visit eat lunch on the shore, search out wildflowers or simply admire the mountain panorama surrounding the lake. The Elbow Lake campground seems luxurious for the backcountry, and relatively large with 15 tent sites. It boasts many amenities — free firewood and firepits, animal-proof pack-rack and food lockers. Most wonderfully, it has a newly-built composting toilet on a scenic platform high above the lake in the woods.

Officially, this campground is one of six backcountry facilities in Peter Lougheed Provincial Park, west of the Elbow watershed. The much more extensive Kananaskis Country, which does include much of the Elbow watershed, was created in 1977 using the Alberta Heritage Fund to develop its 4,250 square kilometres of mountain and

Piping Pika: Rock Rabbit at Risk?

Recent research in the Rocky Mountains has shown that the little pika, also called coney or rock rabbit, is particularly susceptible to warming temperatures. These alpine denizens require cool, relatively moist climates for survival, as they do not hibernate and retain their heavy fur year-round. They survive the grim winters eating from haypiles of alpine vegetation collected during the brief growing season, and spend their entire lives living within an area of less than one kilometre in radius.

In the Elbow watershed, they typically live above 2,100 m elevation, where temperatures never rise above 25C. If the pikas' environment undergoes warming, they must either move to higher elevations in alpine "islands" which become more and more limited in area, or try to move to more northern latitudes. Crossing the warmer valleys between mountains is a significant risk. Researchers have found pika populations in the Yukon Territory decimated by up to 80 percent following a particularly warm or wet winter. In some regions of the United States, pikas have completely disappeared. The pika is considered by many to be the canary-in-the-coal-mine for global warming. The climate in this region has moderated slightly within the last hundred years, and the little pika may indeed be at risk here.

Ka-na-nas-kis? Pardon Me?

The name Kananaskis was used by Captain John Palliser for the river, valley and lakes in this area, derived from a Cree word *Kin-e-ah-kis* (meaning, one who is grateful). The grateful one is said to be a warrior of this name who survived an axe blow to the head during a fight near the confluence of two rivers, later called the Bow and the Kananaskis. It is not the Cree, however, but the Stoney/Nakoda, Blood, Siksika and Kootenai First Nations that have traditional lands in this area of Alberta.

---

foothills terrain. Operation of K-Country and of the campgrounds then shifted from the Alberta Forest Service to park rangers.

While administratively the Elbow Lake and its campground belong to the Kananaskis valley, physically they are part of the Elbow watershed and are focussed downvalley to the east. As such, they logically belong within the Elbow-Sheep Wildland Provincial Park and wildlife reserve, on whose border they are located. No matter, it is one example of many where administration does not conform to the physical watershed boundaries.

The Elbow-Sheep Park, relatively recently established within Kananaskis Country in 1996, covers a large area (over 800 square kilometres) of diverse and largely undeveloped alpine and subalpine habitats in the upper Elbow watershed — truly wild land. No motorized vehicles, no oil and gas, mining or commercial forestry activities and no cattle grazing are allowed within the park. This enhances the backcountry recreation experience, but even more importantly, it protects the alpine and subalpine vegetation, wildlife and landscape of the headwaters of the Elbow watershed — something absolutely critical to the health of this area, as well as that of the lower watershed. Score one for the Elbow!

As the evening cools with the sunset, we move to the common picnic and firepit area to make a warming, mosquito-deterring campfire after our successful source-of-the-Elbow hike.

Chapter 2.

Moving Down the Mountain Stream

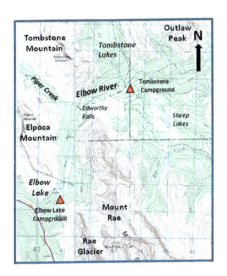

Map of the area downstream from the Elbow Lake Campground

It was the unusual warmth that drew us out of our tents the next morning. We were all used to donning layered clothing, hats and boots in early mountain mornings before heading to the fire pit to make coffee and warm up for breakfast. But this morning was different with a pleasant temperature of about 10C and a light breeze to keep the mosquitoes in the bushes. The clear blue Alberta sky seen above the pink-tinged mountain peaks surrounding the lake

Water Coming and Going: The Hydrologic Cycle

The amount of water on the Earth's surface is relatively constant, and moves continuously from the oceans (holding 97 percent of the planet's water) to the atmosphere, to the land and back to the oceans in what is graphically termed the hydrologic cycle.

Water evaporated from the surface of the oceans is moved by air circulation over the land where it falls as precipitation. This precipitation can then be intercepted in several ways. It can be absorbed by the ground as soil moisture or groundwater; it can run directly into watercourses; it can be evaporated right back into the atmosphere; it can be absorbed by vegetation; and it can be temporarily stored by lakes, aquifers, glaciers, ice caps or snowfields. Eventually, though, it is returned to the giant reservoir of the oceans and the cycle begins again.

promised a fine day of hiking ahead. No rain (or snow!) was likely, unlike the June downpours our hiking group had waded through the previous month; happily that part of the hydrologic cycle would be absent. Instead, the hydrology we were interested in involved the surface waters of the lake and the Elbow River, and the contents of our water bottles.

Mountain-sourced rivers in this region, like the Elbow, get most of their water directly from mountain snowpack melt (spring and summer) and from precipitation in the high foothills (May through July), with the remainder coming from groundwater, other snowmelt and alpine glaciers. Here in the rocky upper Elbow much of the meltwater and precipitation flows directly into the river, but below in the subalpine and boreal zones, where annual precipitation is the highest in the watershed, a considerable amount is intercepted by coniferous forests, absorbed by their dense foliage and the significant water-holding capacity of the needles. The remainder falls through to the ground surface where it soaks into the forest soil.

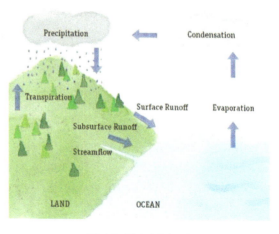

Schematic of the hydrologic cycle

The aspen forests and grasslands lower in the watershed intercept less than the coniferous trees, but still provide the vegetative cover which is critical to the health of the watershed. Vegetation protects the soil from direct raindrop impact, holds soil moisture with its root structure and increases the permeability of the soil, thus reducing runoff and erosion. In urban areas within the watershed, however, like in the City of Calgary, much less precipitation is captured and stored, since more runs directly off into the river over impermeable pavements and roofs. Eventually though, no matter what its source, the Elbow's water is on a long trip to the Atlantic Ocean — an eastward journey through the Bow, South Saskatchewan and Saskatchewan rivers, Lake Winnipeg, the Nelson River and Hudson Bay.

All this seems far removed from Elbow Lake and the pretty summer day taking shape. Our plan for today is to explore the part of the watershed just below Elbow Lake, following the Elbow River down through subalpine forests and meadows to the intriguingly named Desolation Flat and to its confluences with Piper and Tombstone creeks. We boil water for coffee and oatmeal, and treat a few litres of water for drinking during the coming hot day. Then, properly sunscreened, off we set. Skirting the northwest shore of the lake, the trail winds between slippery talus slopes extending far up the mountain side, and the clear blue lake. Already, a couple of fishermen are standing at the water's edge, casting into the lake for brook trout, and a lone mountain biker shoots past us on her way north.

Past the lake and now on the Big Elbow trail leading north and then east, we walk about three kilometres down this wide, flat-bottomed valley to get beyond the east slopes of Elpoca Mountain. We are following the small meandering Elbow River, stepping across it in several places, because it is just a stream at this point. After about a kilometre, the trail stays on its right bank with the river moving in and out of sight as it follows the terrain downward to the north. In places I can see that the river disappears into a small, more deeply

George and the Missus

George Pocaterra at his cabin (Glenbow Archives NA-695-80)

George Pocaterra was an intriguing character in the early history of exploration in the Kananaskis valley just west of this watershed. Born in Italy, after attending university in Switzerland, he emigrated to western Canada in 1903 with $3.75 in his pocket. In 1905, he established the Buffalo Head Ranch in the Highwood River area, and on horseback, often with his First Nations friends, explored, trapped and prospected for coal in the Kananaskis valley. Concerned that much of the region he was exploring was not officially surveyed, he wrote to the federal government explaining the need. Amazingly, government surveyors soon appeared and carried out the task. It was not until a few years after the surveys that George chanced to discover that they had named several local features for him: Elpoca Mountain, Pocaterra Creek, Pocaterra Valley and Pocaterra Ridge. ➔

incised gorge; in other areas, it occupies a broader valley and is easily seen from the trail.

Rivers obviously increase in size downstream. The Elbow just below Rae Glacier is a tiny stream compared to its size when it meets the Bow River in Calgary. Any tributary flowing into the river increases the area of water catchment and thus the discharge, or water flowing into and within the river. Water Survey of Canada sampling locations along the Elbow illustrate this simple relationship. The lowest flow (96 litres per second) is found farthest upstream on the Little Elbow River where the catchment area is small (129 square kilometres); a much higher flow (255 litres per second) is measured at Sarcee Bridge (just above the Glenmore Dam in Calgary), where the catchment area is 1,190 square kilometres: ten times larger than at the Little Elbow.

Profile of the Elbow River from source to mouth

To accommodate the increased downstream flow, the river channel grows wider and deeper by eroding both vertically and horizontally. The profile of a river along its length is typically concave-upward, with the steepest gradients (and narrowest channels) in the upper reaches and the shallowest gradients (and widest channels) near the mouth. This is indeed the case for the Elbow River, as seen in the hockey-stick profile showing elevation versus distance. Overall, the Elbow's profile is much steeper than that of its neighbouring rivers like the Bow. Near its source, a river like the Elbow flows fast and straight downhill just as we saw in the little stream flowing down from Rae Glacier to Elbow Lake. Further downstream where the gradient is lower (as noted below Elbow Lake in several locations) its velocity slows and the river channel begins to curve and meander through a much broader valley. Thus, the water responds to the terrain and vice versa.

→ Much later, in 1936, on a trip to Italy to settle some family business this Italian-Canadian rancher and mountain explorer met and married Norma Piper, a young coloratura soprano from Calgary studying in Milan. George managed her successful operatic career in Europe until they fled back to Calgary from Switzerland in 1939 at the outbreak of World War II. From the Valnorma Ranch in the Ghost River area, he continued his largely unsuccessful mining and prospecting activities, while Norma performed and taught singing at Mount Royal College in Calgary.

The Elbow River valley, with Elpoca Mountain (L) and Tombstone Mountain (R)

This morning we are walking downvalley on the Big Elbow Trail, a gravel fire road which extends all the way from the lake area to the Little Elbow Campground, nearly 30 kilometres to the northeast. In fact, I had not realized this before, but one could drive along the Elbow all the way from Elbow Lake to its mouth in the City of Calgary, if one had the right permits and vehicle! Walking down this road offers expansive views, changing scenery, appearance and disappearance of the river; the route traces the most direct west-east passage from the Highwood Pass in the Kananaskis valley out of the mountains through to the foothills. Old forestry photographs show the landscape in this part of the upper Elbow valley in 1910–16, when work crews were clearing trails through the subalpine vegetation. A photograph of the trail crew shows nine weathered men in suspendered overalls and sweaty hats, holding the tools of their trade (axe and pickaxe) and staring amusedly into the camera, alongside their cook, obvious in his apron and a necktie. They look like they are proud of having done a good day's work. From other photographs (and from standing in the same location today), it is easy to see why this valley would have been chosen: a wide, relatively smooth valley bottom leading gently down through the towering mountains.

Elbow Trail crew, 1916
(courtesy Alberta Forest History Photograph Collection, Forest History Association of Alberta)

I enjoy walking through the subalpine ecozone almost as much as the alpine. The subalpine covers about one-quarter of the

Elbow watershed, extending from the tree line (at about 2,100 m) at the lower end of the alpine in the Front Ranges down to the highest-growing aspen (at about 1,500 m) at the top of the foothills boreal zone. The vegetative cover varies widely, ranging from thick subalpine forests to dense shrubby growth to meadows covered with a rainbow sea of wildflowers. In this part of the valley, coniferous forests climb up the moderate valley slopes to the upper tree limit. At higher elevations, the trees become islands within bushy meadows and decrease in height; near their highest extent, the trees, called kruppelholz, are environmentally (not genetically, as with krummholz) stunted, growing close to the ground and in lee areas, where they are better protected from fierce winter winds roaring through the valley. Engelmann spruce, subalpine fir and lodgepole pine form the forest cover, and in a few places, open stands of alpine larch are present. Low-growing Labrador tea, juniper, grouseberry, crowberry, Canada buffaloberry and honeysuckle bushes thickly cover the ground: the names make me want to munch on some of them. Interspersed are open meadows rich with wildflowers and grasses. The subalpine meadows around us are a sea of colour on this sunny July day: yellow columbine, red paintbrush, pink wintergreen, purple harebell. At this time of year, the meadows are a feast for the eye, and the photographer!

Common harebell

Here in the subalpine, the climate is cool and damp on average. More snow

The Grizz: Hey, I'm Just Trying to Make a Living Here

These imposing 300-kilogram bears with their dish face, distinguishing shoulder hump and streaky fawn-coloured fur used to roam widely through diverse habitats, even out on the prairies and as far south as Mexico. In the local Alberta foothills, "rancher control" and rabies reduced populations drastically in the 1940s and 1950s. Today, grizzlies are found only in the mountains, foothills and boreal habitats of western Canada, the northwest United States and the Arctic, subjected to increasing human-caused mortality (both regulated and illegal hunting, vehicle and train collisions, self-defence, relocation of problem animals and malnutrition due to habitat alterations). Only a few live in the Elbow watershed.

The fate of the grizzly bear in Alberta is becoming increasingly uncertain and the grizzly was designated a threatened species by the province in 2010. The total Alberta population is fewer than 700 bears (down from an estimated 6,000 in pre-European times) while in neighbouring British Columbia the current population is as high as 17,000. Over a decade ago, the Eastern Slopes Grizzly Bear Project estimated that only 38 to 50 grizzlies inhabited all of Kananaskis Country and the adjacent Bow-Crow Forest. →

---

accumulates here than in any other zone. Although there is no weather station in either the subalpine or the alpine zone in the Elbow watershed, readings from the nearby Kananaskis Pocaterra station show that the majority of precipitation in this area falls in May and June, with the heavy snow season lasting from December through April. The daily average temperature for the year is only 0.9C (compared to 4.1C in nearby Calgary). That difference may not seem like much, but it is similar to the temperature difference between Calgary and Fort McMurray (800 kilometres to the north). Only June, July and August have average daily minimum temperatures above freezing (a total of only 96 days a year compared with 170 days in Calgary)! It is a severe climate indeed. But because the landscape in this broad Elbow valley is relatively undisturbed, the intact vegetation is able to hold the snow and rain, deferring its ultimate rush downslope into the creeks and the main river and thus spreading out the water received in the lower watershed over more months of the year.

Now I get out my hiking book and map again, as we have planned to walk up to Piper Pass (appropriately named after Norma Piper Pocaterra) between Elpoca and Tombstone mountains. I think we have reached the ill-defined intersection with a smaller trail that should lead down to and across the Elbow, and then up into the valley along Piper Creek. None of us is certain that this is the place, but we are all eager to see what Rockies hiking guru Gillean Daffern describes as one of the most beautiful alpine valleys in the eastern slopes on the way to Piper Pass.

Apparently, this old pack trail leading through Piper Pass north around Tombstone Mountain must have hosted packhorses with better luck finding and following the trail and fording the Elbow. After a lengthy discussion over the map, our group sets off on the leg-scratching shrubby track which, in retrospect, may have been just a game trail, and work our sweaty way steeply down to the river. Here the water is about 60 cm deep and two metres across, so we won't be stepping neatly across it as we did further up the trail. Dropping a pack, one of us struggles upstream to find a ford to a trail on the opposite bank, and another grimly works her way downstream for the same purpose. The others stay put, me puzzling over the annoyingly uninstructive map. No luck on any account. We give up, apparently not determined enough to see the valley or to get our boots wet. Save that one for another day — time for lunch, a solution that always raises our spirits.

While we were rooting around among the willow and cow parsnip, we were also on special alert for others sharing our space — namely the few remaining grizzly bears that frequent this subalpine area. To many, the omnivorous *Ursus arctos* is the ultimate wild animal, with close encounters rigorously avoided.

Being bear-aware is critical when in bear territory. It means hiking in groups of six or

→ Grizzlies hibernate during the winter months but are out hunting, scavenging and raising their young the rest of the year. They are typically found at lower elevations in spring, moving higher as the growing season progresses, until they den for the winter at high elevations (2,000 to 2,300 m) on steep north- or east-facing slopes, where snow cover is significant. While there are unfortunately fewer grizzlies in these mountain and foothills habitats every year, due to encroaching human activity and the bear deaths that result, nevertheless it is their territory and people must stay bear-aware.

In the Elbow watershed, as in adjacent watersheds, grizzly habitat has been significantly altered over the past hundred years. Roads, tracks and trails for industrial, commercial and recreational uses have reduced and fragmented grizzly food sources and corridors. Lower in the watershed, the proliferation of acreages and agricultural operations have increased the incidence of conflicts, and forest clear-cutting and fire suppression have reduced the quality and quantity of bear habitat. In the adjacent Bow Valley, studies have shown grizzly reproductive rates to be the lowest in North America. As a result, the Alberta government has taken some steps to help the population recover and become sustainable, including a 2006 three-year moratorium on grizzly bear hunting, which has been renewed annually since 2009.

more if possible, and constantly looking for bear sign: tracks, scat, ripped-up stumps or logs, turned-over rocks or other signs of digging for tubers or roots. Where ravens are seen circling overhead or the smell of rotting meat is obvious, bears may be feeding nearby on a recently killed animal carcass. Bears particularly like to forage in areas with juicy berries (blueberries, buffaloberries, grouseberries), cow parsnip found by creeks or rivers, and patches of sweetvetch, or they will dig for ground squirrels or marmots. Making noise, particularly in these bear-feeding areas, is a good way to make bears aware that people are in the area, so we hikers are used to talking and calling out "Yo-bear" at frequent intervals.

Back at Elbow Lake, we and the other campers have been very careful with all food and cosmetic products — anything with attractive odours. Cooking is done at a safe distance from the tents, and all attractants are put in bear-proof lockers, or hung from the pack-rack, when not in use — especially at night when bears and other foragers are most active. Everything associated with food and cooking is cleaned carefully and slops are burned or put down the outhouse. So far, these procedures have worked — no bear encounters. However, a week after we had returned home from this camping trip, we heard unsurprising reports that grizzly bears were frequenting the area around the lake and Elbow Pass, an important corridor for bear movement, particularly in the pre-berry season. Official bear warnings were issued by Park personnel and one trail was closed to allow a grizzly to go about its business undisturbed.

Of course, grizzly bears are not the only mammals living in this subalpine zone. Black bears, bighorn sheep, Richardson's water voles, golden-mantled ground squirrels, and short-tailed weasels are frequently seen here. Less frequently seen are cougars, white-tailed deer and snowshoe hares. The only wildlife we see, however, are ground squirrels and two Clark's nutcrackers, also known as camp robbers. After some discussion and map consultation, we decide to head for Tombstone Lakes in lieu of Piper Pass — only another four kilometres down the trail. In this heat, the idea of a lake seems very attractive.

View downvalley toward Tombstone Mountain

Back in the sunshine, we head toward our interim destination, Desolation Flat, so named in 1919 by George Edworthy (son of the well-known Calgary pioneer Thomas Edworthy). From the trail here, the river itself cannot be seen, hidden as it is in its gorge bordered thickly by spruce, pine and willow. It runs swiftly downhill toward Edworthy Falls, larger than it was at its exit from Elbow Lake since it has been joined by several tributaries flowing from the slopes of Mount Rae. Its course from Elbow Lake to Edworthy Falls is the second steepest in its entire length, with a gradient of 54 m/km. On a horseback trip to further explore the upper Elbow, the ever-adventurous Edworthys (George and his wife Myrle) found two waterfalls on the Elbow here, which they named Desolation Falls; they were later officially renamed Edworthy Falls in their honour. Trudging along in the quiet, heat-lazy afternoon, we all perk up at the sound of tumbling water and detour over to the edge of the gorge. Below, frothy water foams over and down a series of rock steps to the next level of the sun-dappled river; we feel cooler just watching the spray from its constant plunging descent.

Edworthy Falls on the Elbow River

Looming above our little group to the north is the distinctive bulk of Tombstone Mountain. This two-headed limestone giant — it has both a north and a south summit — was named in 1884 by Dr. George Dawson (that inveterate mountain-namer) for the giant grey rock slabs near the top which looked to him, and to us, like tombstones. The mountain has subsequently lent its name to the Tombstone campground nearby, the lakes that are our destination, and a jewel of a pass between it and the west slopes of the Banded Group.

Today, Desolation Flat appears anything but. Here the valley has widened considerably, and the view from the trail is of a broad expanse of pleasing subalpine plain centred among the heights of Tombstone Mountain, Mount Rae, the Banded Group and Cougar Mountain. Where the trail descends sharply to the valley bottom, the winding Elbow is once again visible, as is its confluence with Tombstone Creek burbling down from the pass, and the creeks descending from Lake Rae and Sheep Lakes to the south, giving a good sense of the breadth of the watershed at this point. In fact, in an interesting twist, the Sheep Lakes creeks used to flow to the south, instead of north. Really?

Rivers can be voracious, expanding their drainage area by capturing drainage area from other watersheds, and the amiable little Elbow has committed just such a nasty act of piracy. Where watersheds abut at their drainage divides, headward erosion by the more vigorous tributary can eat into the less actively eroding side of the divide and thereby divert water from that drainage basin into its own. Such active headward erosion will be more vigorous if the stream, most often near the headwaters of the watershed, has a steeper gradient or has less resistant rock to erode.

In the 1940s, geologists from the Alberta Research Council concluded that the Elbow River had captured the upper part of the Sheep River drainage in pre-glacial time through headward erosion. Recent research by geographers at the University of Calgary has led to the same conclusion. The drainage divide which now lies in the vicinity of Sheep Lakes was previously much farther north and west, and the loss of drainage area by the Sheep is the Elbow River's gain. And here in Desolation Flat, the Elbow here looks worthy of the other names given by members of the Palliser Expedition: Swift Creek and Strong Current. In fact, we are glad to see a bridge to cross the now six-metre-wide, albeit shallow, Elbow!

Lower Tombstone Lake and the snowfield above (Photo courtesy of Guy Kerr)

Moving Down the Mountain Stream

Tombstone Backcountry Campground, a small tenting and equestrian facility, is sited centrally in this plain, amongst the trees on the left bank of the river. When we reach the river near the campground, everyone gladly soaks neckerchiefs and hats in the icy water. We then trudge up the Tombstone Pass trail on our way to our revised destination for the day — Tombstone Lakes — happy to be on the last outbound leg of the hike.

The two Tombstone Lakes are found on the east flank of Tombstone Mountain, just below Tombstone Pass, less than two kilometres from, and about 200 metres above, Tombstone campground. These small tarns are nestled in a cirque amongst the subalpine spruce and fir trees, just below a small snowfield on the mountainside. Gillian Daffern describes them in her guidebook as "staggeringly beautiful: silky dark green water backdropped by Tombstone South and Tombstone Mountain" and they fully live up to this billing.

Perhaps even more than the Rae Glacier cirque at the Elbow's headwaters, this cirque is striking — over a kilometre in diameter, with a small snowfield on its slopes and a steep headwall extending up over 700 metres. The two aquamarine lakes, each dammed at their east end by a recessional glacial moraine, sparkle in the sun and provide cool respite. A lone fisherman on the south shore of the lower tarn casts for stocked cutthroat trout. We plunk ourselves down at the water's edge, taking our boots off to soak our feet. Out come the cameras again, but photographs won't quite capture the essence of this tranquil spot.

Twenty minutes later, we reluctantly retrace our steps to leave the cirque and head home. As we emerge from the trees on the mountainside, I look left up to the lush alpine meadows of Tombstone Pass and at Tombstone Creek rushing down to the "Big" Elbow, the river we have been following all day. On the far side of the pass, water flows north to the main Elbow's smaller sister, the Little Elbow River. The water will all eventually end up in one river further downstream, but the pass represents an interior drainage divide in the Elbow's watershed. We agree to return another day to explore its other side.

Back down at the Elbow, I filter enough water for the long return hike, and we begin the hot, dusty trek back to camp in the waning afternoon sun, singing lustily for bears as we go. By the time we reach camp, we are all satisfied with today's adventures — more explorations in the Elbow's headwaters, stories of piracy and love affairs, and the magic of water in its various river, tarn and waterfall forms, and in our icy water bottles.

Chapter 3.

Where Mountains Meet Foothills

Map of the Elbow Loop, the trail around the Banded Group

There was a hum of anticipation in the car approaching the Banded Group – the four mountains which form an imposing phalanx on the very eastern edge of the Front Ranges within the Elbow watershed. It is these four limestone behemoths — Banded Peak, Outlaw Peak, Mount Cornwall and Mount Glasgow — that we five hikers have decided to circumnavigate in our first backpacking trip together. At the end of Highway 66 in the Little Elbow campground, we help each other heft heavy packs onto our backs, ready to set out on the Elbow Loop in the cool, cloudy July morning.

The Elbow Loop is a 45-kilometre route of gravel fire roads and narrow trails that encircles the Banded Group. A favourite novice-to-intermediate circuit for mountain bikers, who can complete the loop in about eight hours, it is also a popular equestrian trail, and facilities for outfitting and horse camping are found along the way. For hikers, it is a three- or four-day journey. We have planned two days of 12-kilometre hikes with overnights at the Mount Romulus and Tombstone campgrounds, and a long final day of 20 kilometres to complete the loop back at the Little Elbow campground. Working counterclockwise, we will first be walking west into the mountains away from the foothills, and then circling back along mountain/foothills boundary to finish our trip.

Banded Group-(L-R) Outlaw, Banded Peak, Cornwall, Glasgow

The trip starts well, as we walk five abreast up the modest incline of the wide fire road, periodically adjusting our packs, listening to birdsong in the trees, and admiring the surrounding mountains. We are starting in the northeast corner of the loop, walking up the valley of the Little Elbow River and will cross this main Elbow tributary several times enroute. On our left stands the Banded Group (also

Where Mountains Meet Foothills

known as the Glasgow Group) which lies between the Big (or main) Elbow and Little Elbow rivers. This group of peaks was distinctive enough to attract the attention of the Marquis of Lorne, then governor general of Canada, who sketched them in 1881, and alpinist-artist Edward Whymper, who published an engraving in 1885.

Just beyond the Little Elbow Campground, we pass a trailhead sign and small track leading off to the right upslope into the forest. This is the trail to Nihahi Ridge. Although not venturing in that direction today, we have climbed this favourite ridge on the forefront of the Front Ranges many times in the past. *Nihahi* means rocky or steep cliff mountain in the Stoney language. This massive and amazingly straight, steep rampart stretches 10 kilometres north-south, and is easily identified from the prairies to the east (just north of the distinctive Banded Group, of course). Nihahi provides a graphic introduction to the geology of the Front Ranges, where the so-called McConnell Thrust has pushed older layers of Devonian Palliser formation limestones over the younger Cretaceous sandstones. This is the opposite of normal geologic layering where the youngest rock layers are topmost and the layers are older with depth. Geologists frequently take field trips there to ogle the formations.

The Banded Group: My Horizon Benchmark

The Banded Group are always recognizable on the horizon. With a summit elevation of 2,935 m, three-sided Glasgow has the distinct, classical pyramid structure and was originally, appropriately, called The Pyramid. That name was officially changed in 1922 to Mount Glasgow, in honour of HMS Glasgow, a light cruiser which took part in World War I. At its highest elevations lie massive grey Carboniferous limestones; below them are Devonian/Carboniferous limestone layers and the cliff-forming Palliser formations on its southern and western faces.

Mount Cornwall, with a summit of 2,970 m, earns its place as the highest of the four mountains in the Banded Group by just a few metres. Its upper elevations are erodible limestone layers, with the gun-metal vertical face of Palliser limestones below. Like Glasgow, it was also named for a World War I cruiser, and is easily recognizable from the prairies to the east, but for a different reason. Each winter, deep snow accumulates in a northeast-facing cirque below the summit; this patch is the last visible winter snow to melt in this area, if indeed it melts. While not a pretty mountain, Cornwall is a favourite of adventurous hikers and backpackers. Daffern's hiking guide describes its "gloriously long summit ridge of brilliant orange shales...lined both sides with stupendous cliffs of palest grey." It is typically ascended from its eastern slopes, using South Glasgow Creek as the approach, but the route up the Talus Creek valley is another option. ➔

Nihahi Ridge, viewed from Forgetmenot Ridge

Halfway up the trail at ominously named Bear Meadow, a broad grassy expanse which opens up just above the forested slopes, the geology of Nihahi Ridge can be seen in the slope face ahead. Rising from the meadow and the McConnell Thrust is the straight-sided Palliser limestone, from the top of which the views are spectacular. Among our group, only one of us has climbed above the Palliser because it involves a scramble up a diagonal crack and a traverse across scree to reach the south summit of the ridge. University of Calgary geologists report that the reddish Banff siltstone contains fossils: brachiopods, bryozoans, corals and sharks' teeth. A ridge walk of about nine kilometres is a further reward, if you like that sort of thing. But that is not on the program for today, and so we continue along the Little Elbow Trail.

One of my favourite mountains now lies immediately to our left. This is Mount Glasgow, the northernmost of the Banded Group.

Where Mountains Meet Foothills

I can always recognize it from far far away, even from my house. Up close it is maybe not so distinct from all the other mountains, but that is how it is with mountains. Seen from the trail, the rough northern aspect of Glasgow does not appear welcoming. For good reason, climbers typically approach Glasgow from the friendlier south, often in conjunction with its southern neighbour, Mount Cornwall. There are also popular scrambling routes up its east and west ridges. The western and northern flanks of Mount Glasgow are skirted by the Little Elbow River on its way east to its junction with the main Elbow. Several unnamed creeks rush down these slopes to meet the Little Elbow and add to its volume.

We stop for an early lunch at the side of the road, gratefully dropping our backpacks onto the grass. This action makes me feel like I am so light I could fly. As we perch on roadside rocks, a string of a dozen horses and riders headed for the equestrian campground plods slowly past. The mounts of the less experienced riders skitter and dance nervously past us, confused by five figures sitting by the track. Sitting at hoof level, we tense, ready to leap out of the way of a kick, but eventually all pass without incident, though not without our envy regarding the mode of transportation.

The wildflowers that are so numerous along the trail at this time of year require many stops. The photographer's challenge is to squat down far enough to zoom in on

→ Outlaw Peak is the smallest (at 2,850 m) of the Banded Group, and the most recently named (albeit still unofficially). In 1974, climber and mountaineer Don Forest thought he was being clever in calling the mountain Outlaw — as a companion to its neighbour, which he thought was called "Bandit."

Banded Peak's pleasing form is dominated by its dove-grey limestone/dolostone triangular summit and signature dark charcoal band of steep Pekisko limestone cliffs below. This band, for which it was named about 1896, is especially noticeable when the mountain is snow-covered (since snow does not accumulate on the steep cliffs). Banded Peak is often climbed in conjunction with the other three peaks as a traverse, or alternatively on its own: a gruelling ascent of 1,400 m to the summit. Both early and late in the season, while surrounding peaks are being rained on, these mountains are covered with fresh white snow for some obscure meteorological reason. All in all, this is an island of four impressive mountains in the middle of the watershed.

The Riparian Zone: The Obvious Part

Most people do not give riparian zones much thought, but they constitute a critical part of the ecosystem of a river. And in Alberta their health is vitally important, as riparian areas support 80 percent of our wildlife. Riparian vegetative cover protects streambanks from erosion by stabilizing the soil with its roots, and provides food, cover and connected travel corridors for many species of birds, mammals and other animals. When an unexpected moose or bear suddenly shows up in the suburbs of a city like Calgary, it is usually because they have followed the continuous riparian zone of a river into the suburban area.

Up here in the mountains close to the rocky glacial source of the river, the vegetation of the riparian zone is significant for the life of the river. It provides organic material (leaf litter and woody debris) to the water, which is required for the food chain. Its cover gives shade for fish and other aquatic species within the river. It also buffers the severity of seasonal flooding by soaking up excess water and by storing water; as a result, peak flows moving downstream are reduced, as is the velocity of flow through friction. This flooding is also important for the riparian vegetation; it brings nutrients and moisture for growth. ➔

---

a flower at the right angle while wearing a heavy backpack. Our smallest member "turtles" on the trail — lying flat on her backpack with arms and legs flailing in the air. We eventually have to help her up — trying not to laugh too hard.

It is an easy day in terms of elevation gain, as our destination — the Mount Romulus backcountry campground — is only 170 metres higher than our starting point in the Little Elbow campground. But the last few of the 12.5 kilometres seem long indeed, and we are grateful when the campground sign appears suddenly beside the fire road. It is about three o'clock in the afternoon, and pitching camp seems like a wondrous thing (like that floaty-flying feeling when the backpack comes off). In our search for the tent sites, we startle a naked woman taking an open-air shower with her (friendly) dog standing guard, in what seems to be a semi-permanent part of the equestrian camp. She wetly points us in the right direction and at long last, we locate the seven backpackers' campsites a little further down the road.

The campground is located between Mounts Romulus and Glasgow, where the North, South and West Fork creeks join to form the Little Elbow River. It is a moderately sized facility with 14 tent sites, but has prized amenities like picnic tables, water, toilets, food lockers and firewood. It seems like a luxury hotel to us, within hearing distance of South Fork creek babbling beside the trail. The campsite is named for Mount

Romulus, the blocky flat-topped mountain with steep east-facing cliffs standing tall (at 2,832 m) on the north side of the Little Elbow. Romulus, recognized by the white snow cornice stripe across its broad limestone summit, and its slightly higher brother mountain, Mount Remus, were named in 1940 for the famous twins who, according to legend, founded Rome.

→ A healthy riparian zone, as is generally found along the Elbow River above Bragg Creek, filters contaminants from any upland runoff. Although contaminants are not usually an issue this far into the mountains, this filtering function can be important in agricultural areas (for example, farther downstream on the Elbow).

Besides hikers, frequent visitors to the riparian zone in the mountains include grizzly and black bears, elk, moose, deer, water voles, chipmunks, cougars, wolves, and coyotes. Beavers and water shrews live farther downstream from this subalpine location. Upland birds and waterfowl, including Clark's nutcrackers, swallows, chickadees, warblers, sparrows and kingfishers, plus ducks and snipe, also use this zone for food, shelter and water.

South Fork creek and its riparian zone, with Mount Romulus in background

Before dinner we explore the area, starting with the South Fork, an icy, clear and shallow creek about three metres wide, rushing over limestone cobbles and boulders. From our campsite, we reach the water by walking

The Aquatic Zone: The Watery Part

The nature and diversity of the aquatic zone varies widely from river source to mouth, in large measure dependent upon the availability of organic matter as a basis for the food chain. Up in the mountains, little organic material is present in the river, the only source being the leaves, twigs, bud scales, seeds and flowers dropped into the flowing water. Farther downstream, these materials are shredded or collected by aquatic insects (macroinvertebrates or insects without backbones) such as stoneflies, caddisflies and blackflies to produce nutrients in the stream. More sunlight for photosynthesis is available to this wider river, so that aquatic plants (such as mosses and liverworts) and algae, clinging to rocks in the stream, diversify and produce additional organic material.

Once these producers at the bottom of the food chain are established, consumers (like stream bugs that cannot produce their own food) can survive. Toward the mouth of a healthy river, much more organic material is received as sediment from upstream sources and runoff, supporting all levels of the food chain (from fungi, bacteria and plankton, through aquatic insects, algae and fish, to amphibians, reptiles, mammals and birds. And fishermen.)

through its riparian zone — the banks of the stream and its floodplain that act as a transition between upland areas and the water itself. This zone can include wetlands, level uplands or high steep banks sloping down to the river, plus the plants and animals living there. Here at the South Fork, the riparian zone is relatively narrow. Scrubby willow and alder, paintbrush and elephant-head mix with mountain avens, river beauty and various grasses, while small subalpine fir and spruce stand higher. The area is easily disturbed, and in this location several dirt paths have been created by campers walking down to the stream for access to the water. Farther downstream, the deep tracks of horses from the equestrian campsites constantly crossing the stream to head farther west have caused erosion on the banks and throughout the riparian zone.

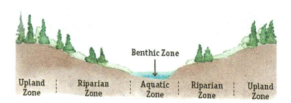

Diagram of four river ecosystem zones

As we idle on the river bank, its constant sound and movement draws our attention to the moving water itself. Besides the riparian zone, the river ecosystem also includes two parts within the stream itself: the aquatic zone and the benthic zone. Playing an important role in the health and diversity of the river ecosystem, the aquatic zone includes the moving water, the living creatures within the water, and floating plants – in other words, everything from the water surface to the river bottom.

The type and quality of the aquatic habitat determines the presence or absence of fish in the river. Water temperature, predators, food sources, spawning/breeding environments and water quality all will affect their location. In the Elbow watershed, rife as it is with cold-water mountain-sourced streams, cold-loving fish species naturally abound. In spring and fall, Alberta Fish and Wildlife also intermittently stock a few ponds and streams within the watershed, including McLean Pond, Talus Lake, and Cougar and Ford creeks. Each year, for example, McLean Pond receives over 7,000 rainbow trout; in 2008, Talus Lake (above the Mount Romulus campsite on Mount Glasgow) was stocked with 300 cutthroat trout fry.

Trout species abound in the watershed. The rainbow trout, originally found only in northern Alberta, is now stocked in many cold-water streams and lakes in southern Alberta. The cutthroat is native to mountain and foothill streams and likes even colder water than the rainbow. Both thrive in the Elbow River, eating flying insects, invertebrates and small fish, and are favourite fly-fishing targets. The bull trout is Alberta's provincial fish, native to all Alberta river systems like the Elbow with headwaters in the Rocky Mountains. Though it has the broadest natural distribution in Alberta of any trout, it is now classed as a vulnerable species on the eastern slopes, as it has been seriously outcompeted for food and habitat by the brook trout (introduced to Alberta waters from eastern Canada in 1903). The bull trout is often erroneously called Dolly Varden, confused with a northern species named for Charles Dickens' character in *Barnaby Rudge*, who wore pink-spotted clothing similar to that fish's colouring. Bull trout spawn in autumn and are typically found in stream pools and backwaters, where they are caught by bears, osprey and fishermen. Brown trout, however, are found further downstream, in the slower streams and warmer waters east of the foothills.

We sit musing by the water with thoughts of fresh-caught trout for dinner, but without licenses and fishing gear, this is just a

The Benthic Zone: The Bottom Part

The benthic zone – the streambed and the plants and animals which live in, under or close to it – naturally plays an important role in the food chain and the health of the river ecosystem. In the Elbow watershed's streams, benthic macroinvertebrates include mayfly, caddisfly, stonefly, damselfly and dragonfly larvae, and worms, all found under the stones of the streambed. These macroinvertebrates particularly love to populate riffles in streams – the areas where the water burbling over gravel and cobbles provides an abundance of oxygen and food particles. Here the river ecosystem is diversified to a point where nutrients, organic matter and invertebrates are present and further diversification can take place with the introduction of vertebrates: the ever-popular fish.

Fish are at the top of the fresh water food chain. These cold-blooded, torpedo-shaped animals are perfectly suited to their life in the aquatic environment. With fins, scales and a "lateral line" system (external pores connected to nerve endings used to detect vibrations and temperature changes), they can move, protect themselves and communicate with their aqueous surroundings. They forage quickly and widely for food, which includes other fish, minnows and fish eggs, plus plankton, insect larvae, worms, aquatic plants, algae, terrestrial insects and crustaceans.

---

mouth-watering fantasy. Instead, we return to our campsite and satisfy our sizeable hiking appetites with something more mundane. When the sun disappears behind Tombstone Mountain and the stars sparkle above, we warm up with fortified tea, then stow the food and cooking gear in the lockers, put out the fire and head off to the tents.

The next day starts slowly — no particular excuse for it other than, eh, it's only 12 kilometres to the next campground. By 11 o'clock (shocking!), we are underway, heading south up the Little Elbow Trail toward Tombstone Pass. After an initial slog up out of the valley bottom, the trail we are following through subalpine forest rises 400 metres to the pass, and gives wonderful views in all directions. On our left are the massive limestone outliers of Mount Cornwall; on the right, Tombstone Mountain's south summit rises impressively high above us.

As our group continues uphill in the heat of the day, we discuss the stories we have heard about icy Talus Lake, now just east of us on Mount Cornwall, but tucked up behind a formidable ridge. Talus Lake is a popular destination for backcountry hikers and fishermen alike. The shimmering turquoise tarn, often stocked with cutthroat trout, is surrounded by shaley scree slopes (*talus*, of course) and alpine meadows in the sheltering cirque — well worth the 500m climb to get there. Some visitors come from Little Elbow campground, first biking 12 kilometres and then hiking five kilometres up the valley to the lakes; others day-hike in from Mount Romulus campground. We agree we will have to attempt that adventure on a subsequent trip.

When we at last reach the fork off the main trail that leads up to Tombstone Lakes, a lunch break seems in order. We head into the shade of the subalpine forest edge and sink gratefully down. A rich, pungent and almost offensive vegetation smell, somewhat like overripe cheese or wet dog, overpowers the delicate fir and spruce scents. We spot telltale clumps of the purplish-white flowers of valerian — or wild heliotrope (although not truly a heliotrope) — with its sturdy stems and dark green leaves. This tall perennial plant has apparently had many practical uses over time: a remedy for insomnia in ancient Greece and Rome; a perfume (hard to imagine!) in the 16th century; food and medicinal uses (for stomach ailments) for the Plains Indians; a current herbal remedy for anxiety and intestinal disorders; and a source for the commercial tranquilizer known as Valium. The smell gets us back on the gravel trail, walking single file to grab whatever shade the road edge has to offer. All morning, we have been following the South Fork of the Little Elbow uphill and for the first time, we cross the narrow creek as it runs almost silently through a brushy flat of low willow and other scrub. One of the group, with her water supply running low, bends down to refill a water bottle for treatment. Forgetting the critical balance conditions imposed by

Aromatic valerian

her tall, heavy backpack, she takes an unplanned header into the icy stream and claims it feels wonderful.

Just one more kilometre and we reach Tombstone Pass, the broad saddle between Tombstone Mountain on the west and Outlaw Peak on the east, and a low point in the divide between the Big and Little Elbow rivers. The South Fork now runs behind us back down to the Little Elbow, and ahead I can see a small creek preceding us to the Big Elbow and our campground destination. The panoramic views in this shrubby col meadow with its extensive sap-green patches of subalpine larch trees are spectacular.

Noting the larch groves, I make a mental note to return in the fall when these deciduous conifers, also called tamaracks, will turn an intense golden yellow before shedding their needles in October. They are equally lovely in spring, when their soft, pale green needles emerge, contrasting with the dark evergreens and remaining snow patches.

Stands of larch in Tombstone Pass

East of us now is Outlaw Peak, the third massive limestone mountain of the Banded Group, and another that we admire from a distance. As we head downhill on the 300-metre descent to the Big Elbow River and the last three kilometres to the Tombstone Campground, expansive vistas of the Elbow valley stretch out below, including the south summit of Tombstone Mountain and Mount Rae. Too soon, though, we are in prime grizzly territory.

The Lovely Larch

These unusual trees grow in relatively harsh mountain environments, often on northern slopes, in rocky or gravelly soils and on talus slopes. Their slow growth as young trees (about 1.5 cm a year) allows their root systems to become established early and reduces topkill from wind or heavy snow. At 25 years and not half-a-metre tall, their solid roots extend an equal distance into the ground.

Larch are important as a food source for the blue grouse (Alberta's largest grouse), which feeds heavily on its needles; as protective cover for birds and animals at high elevations near or at tree line; and as a stabilizer of snow cover on slopes. The latter helps protect the watershed, reducing the incidence and impact of avalanches and associated erosion on otherwise unprotected alpine slopes. The larch has also been used by native peoples for medicinal applications, snow shoes and toboggans, and goose decoys.

Cow parsnip, popular delicacy for bears

The first indicator is a large clump of cow parsnip, then another, and another, on both sides of the trail. This big, bold member of the carrot family is easily recognized by its large flat umbels (flowerheads) composed of little white flowers, its musty odour and its height (up to two metres). It is common in moist, shaded riparian areas in subalpine and montane environments, as well as in disturbed areas like avalanche chutes. Its juices can cause dermatitis in people, but it has been used effectively for food and for medicinal purposes, such as in poultices for bruises or sores, or a gargle for sore throats. A pretty yellow dye can be made from the roots. But the bear-aware know that cow parsnip is a favourite food of grizzly bears in spring to mid-summer (comprising up to 15 percent of their diet) before they switch to berries. The second indicator is a recent deposit of bear scat at the side of the trail. Uh-oh. We promptly consolidate into a group again, and move quickly and noisily past this grizzly foraging area. An hour later we reach the campground intact.

Tombstone Campground is an equestrian and backpacker facility, much like the Mount Romulus campground of the previous night. Slightly smaller, with only 11 tent sites, it nevertheless has the same amenities, including a gurgling silver creek flowing through the campground and down to the Elbow River. The sky has clouded over, with some foreboding of rain. Nevertheless, we indulge in a bracing cup of tea, feet in the stream, even before setting up the tents. This is somewhat familiar territory, as we had hiked through here from the Elbow Lake campground the previous summer, to visit the Tombstone Lakes. The campground is in a broad green valley, with a striking 360-degree view of Tombstone, Elpoca and Cougar mountains, Mount Rae and its tiny glacier, and Outlaw Peak, with the Big Elbow River bubbling along in the valley bottom.

The next morning, we are packed up early (beating yesterday's dismal performance by two hours at least), since this is the third, last and longest day of our trip. We have 20 kilometres to cover today, much of it on a trail above and beside the Elbow River — the second half of the Elbow Loop. About three kilometres east of the campsite, the river has changed character; at the campground, it was meandering through a broad meadow, but now it foams and splashes its way over the boulders of the deep canyon between Outlaw Peak and Cougar Mountain to the south. The steep sides of the ravine, forested near the bottom, rise a thousand metres to the pale limestone summits of those mountains. The day is cool and sunny, and the walking is single-file on the narrow trail which follows the north side of the ravine. Our loads are somewhat lighter, the views up and down the river valley are remarkable, and the silence meditative. We enjoy it and try not to count the slow kilometres.

The Elbow River canyon beside the Big Elbow Trail

Now we turn north with the Elbow and round Banded Peak, the last of the four mountains in the Banded Group and possibly the most recognizable at any time of year. Hiking to its summit can be done, but would be challenging. We are content to plod around it and admire its shapeliness from below.

As we pass Cougar Creek, a stream which joins the Elbow from the south, the Elbow valley begins to open up again. Cougar Mountain, named for the cougars once frequenting its slopes, is left behind, and Threepoint Mountain appears on our right. The Threepoint Range, which also includes Mount Rose and Bluerock Mountain, with its distinctive midnight-blue, east-facing escarpment, forms part of the eastern boundary of the Elbow watershed in this area, along with Cougar Mountain and Mount Burns (named for Senator Patrick Burns, one of the Calgary Stampede's Big Four).

The afternoon sky continues to darken ominously. We pass the Big Elbow backcountry campground (big only in the sense of Big Elbow River, whereas the Little Elbow campground is a big campground on the Little Elbow River), the third such facility on the Elbow Loop. Its six tent sites, situated among the scrubby spruce and pine near the river, inspire thoughts like "Wouldn't it be wonderful to stop and camp here right now?" But on we go to cover the last 10 kilometres of the trip — we are half done (the glass-half-full ruse)! Now the trail has broadened again into a rather boring fire road in the three-kilometre-wide valley. The summit of the giant foothills ridge, Forgetmenot, on the east side of the valley, is obscured by the black clouds scudding along its length. Above Mount Glasgow (now on our left again) and north to Nihahi Ridge, the ebony sky suggests that this will be no ordinary summer cloudburst. As the temperature drops, we pull on rain gear and pack covers.

We are walking along the border between the mighty Front Ranges of the Rocky Mountains to our left and the rolling foothills to our right. The Elbow, following this north-south divide before turning east to meander through the foothills, has spread out over the cobbles of the broad valley bottom, braided in many places where the channel divides and reunites. Pebbled bars exposed by lower water levels at this time of year are strewn with silver-gray logs, tangled flood debris and, on the larger bars, green clumps of moisture-loving plants like scrub willow and squat purple river beauty. The river channels are shallow. The gradient of the river profile has decreased, and although its volume has grown with the addition of many tributaries, its velocity seems to have slowed somewhat.

Our conversation returns to the subject of life in the river itself. How is it affected by natural and human-induced changes of the types we see or hear about all the time? These influences can be flooding, beaver dams, sediment from bank or slope erosion when vegetation is removed, pollution from oil spills or septic fields, or even climate change. We know that those primary producers in the benthic zone at the bottom of the food chain need light and nutrients to live and grow, so sediment-clogged water, for example, would severely impact their growth. Then, consequences are felt right up the food chain by the consumers like protozoa, insects, fish and so on. One University of Calgary study showed that macroinvertebrates in the Elbow River down to the city limits were in pretty good shape a decade ago. However, the effects of what appeared to be small changes, like reduced leaf detritus in the lower watershed due to fewer trees near the riverbanks, had led to significant changes in invertebrate species. I worry about the effects of erosion from forest clear-cuts or ATV-damaged stream crossings.

But for now, one more stream crossing at South Glasgow Creek, and a sprint across the swing bridge at the Little Elbow before the torrential downpour begins. As we leave the Front Ranges and drive through the foothills to Bragg Creek for celebratory burgers and beer, we pass meadows white with mounds of hailstones. The worst of the storm had missed us — an appropriate end to this memorable odyssey in the mountains of the Elbow watershed!

Part 2.

The Foothills Between

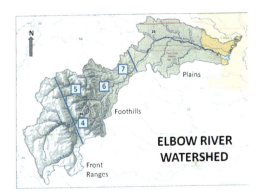

(map courtesy Bow River Basin Council)

Chapter 4.

Traversing the Great Foothills Ridges

Map of the Great Foothills Ridges

My adventurous group of hikers meets weekly to explore the high places that lie west of our homes in Calgary and surrounding region. For a decade, we have scaled peaks and tramped about in the foothills, mainly in this accessible Elbow Valley watershed. The high mountains are exciting, but we return frequently to the Elbow's great foothills ridges which lie between the Front Ranges and the subtler boreal foothills landscape to the east.

This year we have planned a hiking week focused on the Elbow watershed's high foothills. From the Little Elbow Campground at the end of Highway 66 in Kananaskis Country, our daily hikes will cover the great foothills ridges, the mountain outliers of The Family and the Elbow River that connects them. Three long craggy ridges span this watershed from northwest to southeast: Jumpingpound Ridge lies farthest north, separated from central Powderface Ridge by the deeply entrenched Canyon and Prairie creeks, and then, on the south side of the Elbow River, lies Forgetmenot Ridge. The ridges have many similarities, including a relatively uniform elevation (just over 2,200 m) and a length of about eight kilometres. Subalpine forests cover their lower slopes, grassy meadows lie above and extensive patches of bare rock are exposed at their summits. Their gentler west-facing slopes contrast sharply with the steep east-facing sides. Both Powderface and Forgetmenot have resistant limestone outcrops on their east slopes, creating sharp cliffs and long, stomach-grabbing drops from the ridge edges. It is these three ridges, with their marked similarities, yet intriguing differences, that we have set out to explore.

The foothills which lie east of the Front Ranges in the watershed and far beyond to both north and south, are composed of younger, softer rock than the mountains. The relatively soft Mesozoic shales and sandstones of the Foothills region have been continuously eroded since the last major uplift, to produce the rounded rolling ridges of today. These contrast with the higher, more sharply contorted forms of the older limestone Front Ranges just to the west.

Jumpingpound Ridge, our first-day target, was named in 1949 after Jumpingpound Creek, the stream that flows north from the ridge into the Bow River. This creek, just outside the Elbow watershed, was the site of a "jumpingpound," a steep bank near its mouth where buffalo were herded to their deaths by Blackfoot hunters. The ridge is formed of sandstones and siltstones, and even coal at the summit.

Traversing the Great Foothills Ridges

Jumpingpound alpine meadow, with the summit in the distance

A short climb brings a hiker up through the pine-spruce-fir subalpine forest on its west slopes to the alpine meadows, a riot of colour in summer. Purple shooting stars, silky scorpionweed and forgetmenots; pink moss campion and roseroot; yellow buttercups and stonecrop; white mountain avens, daisy fleabane and rock jasmine all bloom in profusion here. Above this on the rocky summit called Jumpingpound Mountain, there is a splendid 360-degree view and more flowers than can be identified in our pack-weight flower books. When we reach the first ridge meadow above the trees on this bright morning, clouds of white butterflies hover above the carpet of wildflowers and grasses, enjoying the hot sun, the wildflower nectars and the freedom of the meadow. And also in the meadow, are three young women with nets, leaping about in hot pursuit of those butterflies. When they take a break, these researchers explain their investigations into this vulnerable species and its changing environment.

Apollo as Canary?

Since 1995, ecologists from the University of Alberta in Edmonton and other research institutes have been investigating the alpine Apollo butterfly — one of the most prevalent butterfly species in the east slopes of the Rockies. This relatively large (up to 75 mm), ironically tail-less swallowtail butterfly, white with black-ringed red spots on its rounded wings and black markings near the wingtips, has very particular environmental requirements for survival. The Jumpingpound Ridge meadows provide them all: the yellow-flowered, fleshy-leaved common stonecrop as its larval host; lots of sunlight and hot summer temperatures; and grassy treeless alpine meadows. In a typical year, about a thousand butterflies are captured, each netted twice over the season. Early in the study, researchers even attached a radar tag the size of a human hair to some individuals, and then followed their movements using a hand-held tracking unit. DNA studies, using a tiny piece of a netted butterfly's wing, are also being carried out to determine the kinship relationships among the populations and subpopulations. ➜

Researcher netting the Apollo butterfly

For over a decade, thousands of Apollo butterflies in each of 17 meadows have been marked and counted four to five times a year to assess changes and movement of individuals within and between meadows. This study provides one concrete example of the environmental impact of a chain of events: warming temperatures leading to constricted alpine meadows in turn leads to decimation of a butterfly population in our upper watershed. Like the pika, here is another canary-in-the-coal-mine. As alpine environments decrease, sensitive populations such as these delicate Apollos have nowhere to go.

Across the deep valley of aptly named Canyon Creek lie the blue-tinged heights of Powderface Ridge, tomorrow's destination in our foothills ridge exploration. Powderface's trails are almost always busy as the ridge is directly accessible from Highway 66. Once called *Thidethaba Baha* (or Mule Deer Buck Hill) by the Stoney people, in 1949 the ridge was officially named for Tom Powderface, a Stoney who lived nearby in the early 1900s. Tom was later said to be one of the philosophers of his nation, and there was great sadness among the Stoney people when he died in the influenza epidemic of 1918–19.

Like Jumpingpound, Powderface Ridge is underlain by sandstones and siltstones, but there are also exposures of Rundle limestone on its eastern slopes. The latter are vertiginously steep, with average slopes of 50 percent, compared to the somewhat gentler average of less than 30 percent on the heavily forested western slopes of the ridge. Rockhounds can find brachiopod fossils in the rock outcrops at the summit. The trail from the highway ascending the south end of the ridge involves 2.5 kilometres of steady, uphill grinding through the subalpine forest until it levels out on the ridge. Several winding kilometres of ridge trail through grassy meadows of wildflowers and stunted kruppelholz, and magnificent views, lie ahead.

→ Over the past few decades the subalpine forest edge here has risen 100 to 200 m in elevation to encroach on the alpine meadows. Whatever the reason — a warming climate, fire suppression, human forestry practices, or even a periodic change not yet understood — the meadow area has decreased by a significant 78 percent since 1952. The encroaching trees have fragmented (or in some cases, eliminated) the meadow habitats of the butterflies. Now the butterflies find it much more difficult, if not impossible, to traverse from one meadow to another to disperse their populations, and this inhibits gene flow. When separated by more than a kilometre of trees, isolated population groups are at risk of extinction.

Elbow River valley, looking southwest from Powderface Ridge

The Elbow valley between Powderface (L) and Forgetmenot (R) ridges, looking east

Perhaps those views keep our hiking group coming back to Powderface. The massive Front Ranges, imposing Nihahi Ridge, grassy Jumpingpound Ridge, bald Moose Mountain and its brother Prairie Mountain, craggy Forgetmenot Ridge, and the broad Elbow River valley stretching east and west — each provides a captivating vista.

South of Powderface, Forgetmenot Ridge is the target of our third day and ridge-climbing quest. Between these two massive ridges flows the serpentine Elbow River in a wide smooth valley, expanded in both width and volume because it has been joined just upstream by the Little Elbow. To get to Forgetmenot, we leave the dusty Little Elbow Campground, cross a swinging bridge over the Little Elbow, trek through the woods and then ford the Big Elbow. Fording the river is relatively easily done late in the summer when water levels are low, but it is an icy experience even on this hot summer day, best done quickly. Then it is just a short jog to the northern base of Forgetmenot.

North end of Forgetmenot Ridge, seen from Nihahi Ridge

Fossil Facts

Stratified rock layers often exhibit remnants or impressions of plants and animals from species mostly extinct today. These fossils were formed by petrification of organisms, by imprints left in mud or sand, by the formation of casts or molds, or even by freezing. Fossils provide evidence of the evolutionary history of life on Earth, using dating methods based on their position in the rock layers and through observation of the life forms identified.

As with Jumpingpound, Forgetmenot is the name of both the ridge and its 2,332 m mountain summit. It was named in 1949 after the tiny periwinkle-coloured alpine forgetmenot which grows in its ridgetop meadows. In fact, it is one of only 21 mountains in the Canadian Rockies named after plants. (Mount Rose is another in this watershed.) The ridge stretches south from the Elbow River to Threepoint Creek, with its mountain summit forming part of the Elbow watershed's southern divide.

Like the other two great ridges, Forgetmenot is underlain by sandstone and siltstone (and some shales). The weathering-resistant Rundle limestone outcrops on its eastern face form even more precipitous cliffs than those found on Powderface. The steep access trails preclude mountain bikers and trail riders from attempting this ridge; it is the exclusive purview of hikers. Main trails to the ridgetop lead up both from the north and southeast. The north access trail has been aptly described as "relentless" — steep, rooty and shaly steps up and up over 500 metres of elevation, through subalpine forest, then scrubbier sparser trees. Finally with a last grind over slippery scree-like debris past the Holey Stone Tree, we reach the grassy meadows at the top and the expanding panorama to the west.

The ridge lies within Don Getty Wildland Provincial Park, a 630 square-kilometre park composed of five discontinuous parcels of land (three in the Elbow watershed) in some of Kananaskis Country's untamed landscapes. Created in 2001, it is named for the 11th premier of Alberta who was also a celebrated quarterback for the Edmonton Eskimos. In order to preserve its natural heritage features while still allowing recreational access, only hiking, horseback riding and mountain biking are permitted in the Park. Forgetmenot Ridge lies in one park parcel that extends west to the Elbow River.

We walk south along the ridge to seek out the features that have enticed us to make the arduous climb up here. Forgetmenot Mountain has been designated an area of provincial significance by the Alberta government, because of a number of exceptional geologic, ecological and wildlife habitat features that remain relatively undisturbed. Of interest to geologists and palaeontologists, and to the scientifically curious, are the brachiopod (lampshell) fossils found in the outcrops below the north summit. Forgetmenot's clam-like brachiopod fossils came from marine invertebrates of the Phylum Brachiopoda, embedded in the rocks that were once part of a vast Paleozoic sea before mountain-building began. So when a hiker discovers a brachiopod on Forgetmenot Ridge, she is seeing a marine creature that lived in a nearby sea around 500 million years ago!

Besides the fossils, there is the Forgetmenot Pot — an 85-metre-deep pothole discovered in 1969 and one of the deepest in Alberta. Potholes, also called pots or sinkholes, are

Periglacial Patterns

Periglacial environments are near-glacial in either condition or location. Here, surface materials may be arranged in symmetrical geometric shapes called patterned ground. Why do these stone stripes, steps, stone polygons and nets of patterned ground develop? Researchers believe they result from freeze-thaw action and other thermal processes.

Felsenmeer (frost-shattered rock), also called blockfields, are typically found in high mountains in the Arctic. In these features, the action of ice and frost has broken the surface rock layer into angular, jagged boulders perched over the rock layers below. Stone polygons show polygonal patterns on the ground surface where stones form boundaries between the polygons. In the stone stripes found on Forgetmenot's west slope, stones form elongated ridges on slopes steeper than those where stone polygons are found.

A rock glacier is also present on the northwest side of the mountain — another reason for Forgetmenot's designation of provincial significance. Rock glaciers are jumbled accumulations of angular rock debris with a glacier-like form (including a steep nose) and can advance like a glacier due to their ice core or interstitial ice (ice in the spaces between rock particles).

> **Curly Sand: All-Around Cool Dude**
>
> The first forest fire observer at the Forgetmenot lookout was George William (Curly) Sand, a tracker, hunter and notorious tale-teller who had come to the Millarville area in 1946 from Minnesota. Hired by the North Fork Stockmen's Association to rid the Sheep River's north fork area of cattle-killing grizzlies, he also worked five seasons at the lookout, likely accompanied by one of his faithful coon hounds. Today, a pretty hiking trail in the Sheep watershed carries his name.

deep vertical or nearly vertical holes usually found in soluble rock (like the Rundle limestone here), formed by the action of running water and open to the surface. It is believed that all such features were formed below the water table, suggesting that at some time in the past, the Forgetmenot water table was considerably higher than it is today. Experienced cavers love these vertical caves with their steep and challenging pitches requiring ropes and ladders to ascend and descend. Fortunately, this pothole's unpublicized location off the beaten track means that hikers like ourselves will not likely run across, or fall into, it by chance.

Evidence of the periglacial (near-glacial) and paleoclimatic history of the area can be found near Old Forgetmenot summit, the actual highest point on the ridge about midway along. Examples of stone polygons, stone stripes, felsenmeer and a rock glacier can all be found on Forgetmenot Ridge. Features such as these have been identified in current and ancient periglacial environments around the world, as well as on the surface of Mars. A connection between Forgetmenot Ridge and Mars? Hard to imagine, yet intriguing. At the summit of Forgetmenot Mountain, evidence of much more recent history is found where a fire lookout formerly stood. The wooden, four-metre-square lookout was built by the Alberta Forest Service in 1954, to serve as the intermediate fire-spotting point between the lookouts at Moose Mountain to the north, and Junction

Mountain to the south, both still in operation. The Forgetmenot lookout was abandoned in 1975; its wooden structure burned down two years later, leaving only the foundation stones to be seen today.

But now dark clouds are gathering on the far side of the Banded Group, a hint that we should soon abandon the ridge. We rush down the slippery shaly track, cross the river and scurry to the swinging bridge as large drops of rain begin to pelt down, driven by the buffeting wind. After reaching the roofed cooking shelter, we glance up at Forgetmenot, now topped with black clouds and blurry with sheets of rain. Good timing!

Why do I like these ridges so much? Part of it must be the ability to quickly reach their summits, to walk about in their flowered grassy meadows, and to enjoy the diverse landscapes seen from their panoramic viewpoints. That's not so easily done with a full-on mountain, and it cannot be done at all in the lower foothills or plains, so these great ridges are prized.

Braided, meandering Elbow River, looking east from Forgetmenot Ridge toward Rainy Summit

For our fourth day, we have planned a different hike — no steep climbs, no high vantage points, but a cool trek for a hot day. We will explore the Elbow itself, between its junction with the Little Elbow and Elbow Falls, about 10 kilometres downstream, on part of the Rainy Creek Loop. This interesting reach of the river includes changing channel patterns over a short distance. In the Front Ranges and foothills areas of the Elbow watershed, we have been able to observe the pattern of the river from most of the high mountains and ridges around it. Looking east down the Elbow valley from Forgetmenot Ridge, for example, there is a clear view of the braided river running through wide cobbly and gravelly flats on its journey downstream. Farther on, northeast of Rainy Summit, the river can be seen to enter a narrower, straighter, deeper channel within a steep valley.

Meanders: The Wandering Watercourse

River channels can be straight (highly unusual), meandering (sinuous), braided (a network of small channels) or anastomosing (multithreaded), or some combination of these. What causes such variations? Usually it depends on the amount and variability of discharge and the erodability of the material of the banks and floor of the channel.

The Elbow River meanders back and forth across its broad floodplain in most of its lower reaches. Why does meandering occur? Virtually no natural watercourses flow in an absolutely straight line. Any small disturbance to the flow of the water will deflect the flow against one bank. The force of the deflected water begins to erode material from that bank; a slight bend in the river develops and the channel gets deeper at that point. The deeper channel and the greater velocity of the water flowing through this bend cause still more erosion. The eroded material from the ***outside*** of this bend is carried downstream and deposited on the ***inside*** of the next bend where the velocity is lower. As this erosion and deposition process continues, the bends become wider and more curved; eventually the bends can even intersect, forming a meander cutoff (oxbow lake).

Traversing the Great Foothills Ridges

Channelized Elbow River, looking southeast past Rainy Summit (R)

Eschewing a start at the jewel-like Forgetmenot Pond near the Little Elbow Campground, we drop a car downstream at Elbow Falls and drive the other back up to Cobble Flats Recreation Area to begin our trek. In the upper reaches of the Elbow and its tributaries, as in most watersheds, the stream channels are straighter, where the water flows rapidly and steeply over, and is controlled by, rock. By the time the young Elbow enters the broader valley near Desolation Flats and the Tombstone campground, its gradient has decreased, its size has increased, its flow has slowed, and the channel has started to meander. Since meander size is directly related to the amount of water in the stream, these initial meanders are relatively small, compared to those found farther downstream.

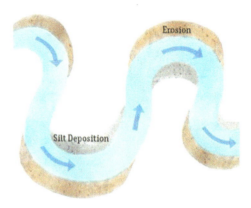

Diagram of meanders in a river, showing areas of erosion and deposition

Cobble Flats, looking west up the Elbow River valley (Mount Glasgow in background)

Today, as we walk on the flats, we have to ford parts of this braided river. A braided river has slightly curved, multiple, shallow and wide channels that repeatedly branch out and reunite, with numerous islands resulting from deposition of its sediment load. Braiding tends to develop where bank materials are sandy or friable, and where bank vegetation is less dense or absent. The bed material is continually shifting, as is the main channel of the river.

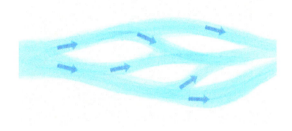

Diagram of a braided stream with multiple channels around islands

In this reach of the Elbow, the channel pattern change seems largely related to geology. East of Forgetmenot and Powderface ridges, the Elbow runs through about six kilometres of Mesozoic sandstone, shale and siltstone rock. These are the more easily weathered and eroded rock formations of the foothills, so the river has been able to broaden its course by eating away at the friable rock, eroding banks and creating many shallow channels and wide expanses of cobbly beach material. Cobble Flats Recreation Area, located on the river about the middle of this section, is aptly named.

As the morning passes, we approach the base of Rainy Summit, just past the confluence with Quirk Creek. Tall spruce trees crowd right down to the river in the steeper valley where now the nature of the channel does change. When the Elbow enters the area of tough Rundle limestone between Quirk Ridge and Rainy Summit, it turns to follow the geologic structure and flows in a straighter, deeper single channel, flanked by heavy spruce forest on steep slopes. There is no room for braiding or meandering here. At one point, these characteristics recommended it as the site of a proposed 40-metre-high "dry dam," part of the government's flood prevention program following the 2013 disastrous floods. Engineering issues soon shelved this idea, however, and currently a dry dam is being considered for one of two sites further downstream.

We are now passing Rainy Summit, the height of land between this part of the Elbow River and Powderface Ridge to the west. This knoll is covered in thick pine and spruce forest, except for its summit and west flank. There, in 1981, crews were clearing brush for the road across the summit, through an area that had been previously logged. As they burned the excess brush, embers blew into the surrounding trees and started a major fire. Snow and wind restricted the firefighting, so that the fire was not under control until the end of that bitter November week. Thirty years later, charred timbers from the fire and wind-blasted tree stumps from the logging remain in a grassy field. That may well have been the most recent forest fire in the Elbow watershed — over three decades ago.

Wild horse on Rainy Summit

When I was hiking on Rainy Summit one fall, three chestnut horses suddenly pranced out of the pine and aspen regrowth. These were part of the herd of wild horses which lives in the watershed, descendents of horses brought to Alberta in the early cowboy days and of horses lost by ranchers or outfitters in the years since. The feral stallions, namesakes of the Mustang Hills (encircled by the Rainy Creek Loop we are on) and the Wildhorse Backcountry Campground farther south in the watershed, stood and stared back at me with ears pricked forward, and then galloped off into the scrub.

Walking on the west bank of the river, we reach the Beaver Flat Campground, nestled on a river flat among the spruce, aspen and willow. With its 55 sites, an interpretive trail leading to a beaver dam and lodge, and basic camping amenities of water, firewood and washrooms, this is a popular destination for tenting, RVing, fishing, and just enjoying the outdoors in the scenic foothills. The 15-metre-wide Elbow flows quickly here and is bounded on its east bank by the steep, spruce-covered slopes of Quirk Ridge. After a brief look around, on we walk, following the gradual descent of the river (only about 100 metres lower than our start at Cobble Flats), past its confluences with Powderface and Prairie creeks — small streams at this time of year — until, at last, we reach the Elbow Falls Recreation Area.

Instead of hopping into the hot car awaiting us, we amble down toward Elbow Falls to enjoy the coolness and the constant slosh of water rushing over rock into deep blue pools below. Accessible year-round, Elbow Falls is located just east of the Winter Gate, which closes Highway 66 into the upper watershed from December 1 to May 14 each year. This closure allows wildlife to move undisturbed to their winter range in the valleys (particularly at Rainy Summit) and gives them quiet for birthing in the spring.

The Elbow valley as a whole is said to be the most visited recreation area in K-Country — even more than the Highway 40 corridor which runs from the TransCanada Highway

Wildhorse Watershed: To Cull or Not To Cull

The Elbow watershed's feral horses live in the Elbow Equine Zone, one of several such zones along the foothills. Feral horses have inspired controversy in southern Alberta since they are seen by many as non-native pests which damage native vegetation, outcompete native species for forage, and steal hay from cattle. Others believe they are naturalizing in their present habitat and are naturally controlled by predators (wolves, cougars) and severe weather.

There are likely fewer than 100 in the Elbow watershed, but they are part of the government-proposed licensed cull in 2014. Culled horses may be adopted, kept by the licensee, or sent for slaughter. The continuing debate over culling is an emotional one in this horse-loving province.

Water versus Land: Who's In Charge Here?

Many geomorphologists believe that running water is the single most important surface process in shaping the Earth's surface. Depending on the velocity of the river, the weather and other conditions, the Elbow and other rivers can erode, carry and deposit a considerable load of rock and other materials, particularly in its upper reaches.

Moving water continually wears away the land below and around it, and carries these materials with it downstream. This load may be moved in many ways: dissolved, floating, suspended, saltating (bouncing along the bottom), or carried by traction (rolling or sliding along the bottom). Upstream, the caliber (size) of the load particles tends to be larger, as slope and velocity are typically greater. Downstream, as the particles abrade against each other, they decrease in size; as the smaller particles move ahead of the larger ones, the load contains finer materials. Along the way, some of the river's load is deposited at the sides or on the bottom of the river, creating islands or bars, or simply raising the bed of the river. Thus, not only the land in the flood plain, but also the river channel and bed, are altered by the running water.

in the Bow Valley, past Nakiska Ski Area, Kananaskis Lodge and up into the Highwood Pass. And this tiny Elbow Falls Recreation Area stands second only to Calgary's artificial Sikome Lake in annual numbers of recreation users. The main attraction, of course, is Elbow Falls, actually a series of small falls and rapids in the river, and one of only two main waterfalls on the Elbow River (the other is Edworthy Falls, farther upstream).

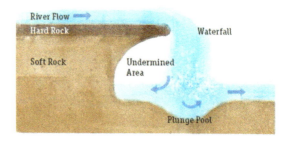

Diagram of the features of a waterfall like Elbow Falls

At and around the main Elbow Falls, where the river rushes over a two-metre drop, it is obvious that the river is hard at work. Waterfalls typically form where there is an abrupt break in elevation in the bed of the river, often where an erosion-resistant cap rock (in this case the Rundle limestone) sits over softer rock below. As water flows over this break, the underlying softer rock wears away, developing into a cave-like undercut below the cap rock. As cap rock inevitably erodes, the falls recede gradually back upstream, leaving a steep-sided gorge downstream. Meanwhile, the force of the water pounding down on the river bed below the ledge leads to the development of a plunge pool; the violent swirling action of the water and scouring of the bottom and sides by in-stream materials continually deepen the pool until it can be as deep as the height of the falls above.

Elbow Falls (Photo courtesy of Robert Lee)

The Falls attract tens of thousands of visitors all year round, to view and photograph the cascade, to picnic or barbeque, and to enjoy the paved interpretive walk around the river and falls. Why are waterfalls so hypnotically attractive? Perhaps it is the combination of the roar of the falling water, the smell and feel of the mist rising from below and the sense of danger when hanging over the protective rail above the rocks and boiling water below. Or maybe it is just the simple beauty of clear water shining over rock. To each his own.

As we wander along the fenced lookouts by the falls, we spot a woman on a stool, hunched over a small canvas in her lap, intent on capturing the essence of the falls before her. Artists have long been fascinated by rivers in general and by waterfalls in particular. Both have been used in drawings and paintings as background for representation of human activities, in foregrounds to illustrate use of the river, and as a means of representing beauty in nature. Early paintings in Canada usually featured rivers in this way.

Early topographical artists came west to map the landscape. They were followed by those who had been hired by the Canadian Pacific Railway to paint tourist posters, and in the first half of the 20th century, by artists instructing in the Banff School of Fine Arts and what is now Alberta College of Art and Design in Calgary. Since the railway, settlement and industry followed the Bow River to Banff and the Rockies, the Bow valley became the focus of an extraordinary body of work. The Elbow River received early attention only around its confluence with the Bow at Fort Calgary, and later in the growing city of Calgary, but not in its upper reaches. Examples of early works include two engravings: *Near Fort Calgary Looking Towards the Rocky Mountains* (Marquis of Lorne, 1882) and *Calgary Bottom, from Frazer's Hill, Looking West* (Thomas Strange, 1882). Other artists — Illingworth Holey Kerr, Jack Rigaux, Curtis Golomb — appear to have discovered the pristine upper Elbow much more recently.

No matter what water's form or location in nature, the Glenbow Museum's Gerald Conaty explains that "the very qualities that make water unique and expressive — movement, change, flow, rhythm …" make artists passionate about the never-ending challenge to "freeze these scenes in time." On that spiritual note, we leave the Falls and head back to the campground to mull over what we have learned about the ridges and river over the last few days. A spectacular day on the Elbow.

Chapter 5.
Meeting the Family

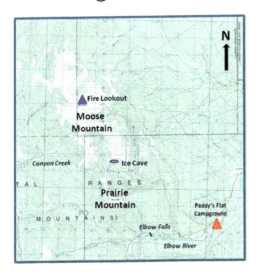

Map of Moose and Prairie mountains, "The Family"

Day five of hiking week in the high foothills of the Elbow watershed dawns bright and clear and very early! We have planned to spend the next two days exploring two mountains east of the great ridges — Moose and Prairie. Standing 200 metres higher than its surrounding topographic features, Moose Mountain is a landmark in this part of the watershed, particularly to First Nations

Mighty Moose and Peaceful Prairie: Mountains of Many Names

The Stoney people called Moose Mountain *Iyarhe Wida*, meaning "mountain by itself" or Island Mountain. Smaller Prairie Mountain to the south was the Younger Brother of Island Mountain. Indeed, the name fits. In 1858, still trying to find the best pass through the Rocky Mountains, members of the Palliser Expedition were headed south from Old Bow Fort on the Bow River to search for a southern route. On the second day of the trip, Captain Blakiston noted in his journal a "marked outlier" which he called The Family due to its peculiar form; this was likely today's Moose Mountain. Further along, he noted a "sharp peak entirely covered with snow" and extremely beautiful; this he named The Pyramid, undoubtedly today's Mount Glasgow. That night, his party camped by the forks of a creek they called Strong Current (the Elbow River), and dined sumptuously on trout from its waters – a fine change from their usual pemmican and tea diet. Thus were Elbow watershed physical features documented in early explorations. →

→ After about 1895, however, Moose Mountain was named after the local denizen and largest deer in the world, *Alces alces* — the ponderous moose — which it apparently resembles when viewed from the north. Prairie was officially named before 1928, either for the extensive grassy meadows on its summit ridge or for the prairie landscapes seen to the east from its summit. Still, perhaps fancifully, the two mountains do form a family — the younger brother Prairie, and the big brother Moose.

peoples. Prairie is shorter in stature and in girth, but has a common ancestry with Moose, both being composed of the same Lower Carboniferous Rundle limestone that is prevalent in the Front Ranges farther west, and seen at Elbow Falls to the south.

Today we tackle Prairie Mountain. Eight kilometres down the road past Elbow Falls, we park in the Beaver Lodge parking lot, and head across the highway to the Prairie Mountain trailhead. The hike to the ridge is a pleasantly challenging one — short (three kilometres) and steep (700 metres in elevation, with an average gradient of 20 degrees).

Meandering Elbow River seen from the Prairie Mountain trail
(Photo courtesy of Robert Lee)

The climb starts right away, and we file slowly up the bank from the highway into the trees on the red-dusty, but often rooty and rocky, trail. After half a kilometre, the trail evens out a bit, feeling like a walk in the park for the next kilometre or so with views through the trees west into the Prairie Creek valley, and south to the main Elbow valley. Other hikers are already descending from the top, and we are passed once or twice by a very fit person jogging to the summit with only a bottle of water as gear.

The absence of mountain bikers and horses is gratefully noted; this unmaintained trail is far too steep for either. After another kilometre of relentless grinding uphill through ever-thinning tree cover, we gain the grassy summit ridge. Ahead in the ever-present wind spreads a long, gently-rising expanse of subalpine meadow leading north to the summit and its rocky cairn. To the west, the meadows slope moderately down toward Prairie Creek, and to the east, the grass ends abruptly in a dizzying drop to trees far below.

View of bald Moose Mountain, looking north from Prairie Mountain

Beside the cairn at the north end of the summit ridge, we look across the Canyon Creek valley to tomorrow's destination — Moose Mountain. Moose fully occupies the view to the north, and from here even the fire lookout on its summit about six kilometres away is visible. With its massive grey dome and multiple crests and valleys, no wonder Moose serves as a landmark among the surrounding forested and grassy ridges and hills.

As the summer afternoon cools off and the wind picks up, whistling through the few stunted spruce at the edges of the meadow, we head back off the ridge toward the shelter of the trees, and the short, steep descent to the valley below. As we reach the final precipitous section of the trail, a steady light rain begins, and we head back to camp to dry out. That evening we wander down the shrubby trail from the campground to Forgetmenot Pond. This popular picnic spot boasts a walking trail around a small, exquisitely turquoise pond, dammed beside the Elbow River and stocked with rainbow trout. A lone fly fisherman across the tranquil pond makes his last casts of the evening, as the pond's waterfowl residents quieten and settle for the night.

Jewel-like Forgetmenot Pond, looking west to the Banded Group

Meeting the Family

The next morning, our sixth and last in the campground, we are up early to head northwest down the road to the Moose Mountain turnoff, near the Paddy's Flat Campground. Our ultimate destination, the great bald pate of the 2,437 m Moose Mountain dome, can be identified from miles around. Its geological features and natural resources, the history of resource development on its slopes, its long-standing pre-eminence as a fire lookout site — all have contributed to the high level of local and regional interest in this mountain.

On our 15-kilometre drive down to the Moose Mountain access road turnoff, after having spent most of a week farther up in the watershed, it becomes obvious that we are leaving the wilderness that we have been enjoying so much. Although still in Kananaskis Country, we are no longer in the wildland provincial park areas of the upper watershed. Instead, this is the less pristine, busier portion of K-Country – the Industrial Development Zone. Here, beef cattle graze placidly beside the road. Clear-cut scars decorate the forested slopes, seismic cutlines criss-cross the landscape, and oil and gas drill pads or wells pop out as dun-coloured squares amid the dark green carpet of trees. Overhead, the helicopters transit back and forth between helipads at oil rigs, the ranger station or forestry sites.

Only four generations ago, in the 1890s, a bustling little Calgary had just been incorporated as a city and was installing its first electric street lights; in southeast Alberta,

The Indomitable Dr. Ings

Out in this Elbow River valley in the 1890s, Dr. George Ings, a well-respected Calgary physician, was indulging his fascination for minerals. On horseback, this amateur geologist regularly set out from the Bragg Creek hamlet area searching the heavily wooded slopes and rocky creek beds for coal outcrops. It was a lonely effort, as there was little activity and few settlers in this part of the rugged wilderness west of the young city. But Ings doggedly believed he would find a mineable coal seam, and indeed he did — not only one, but two.

The first was at the headwaters of Bragg Creek on the west slopes of Moose Mountain. This was exciting, but problematic — in a burst of enthusiasm, he had tons of coal extracted from the outcrop, but then realized that he had no way to transport it to a market. There were no roads in this part of the country — just often-impassable tracks used by the local homesteaders. So the coal sat, stockpiled, in the woods until it was rendered unusable. Not a man to give up on a quest, however, a decade later Dr. Ings followed a track further up the Elbow River, headed up Canyon Creek toward the south flank of Moose Mountain, and, about six kilometres into this deep valley, found a promising coal outcrop on the steep north valley wall. Now, with more substantial transportation infrastructure in place, he was truly in the coal business; he hired a mine manager and the mining work began. ➔

→ Pleased with this success and always ready for a new challenge, Dr. Ings went overseas with a medical corps to serve in the World War I. When he returned at war's end, his geological interest was once again piqued, this time by stories about the Athabasca Tar Sands in northern Alberta, so off north he went and was reportedly the first to take samples to Ottawa for analysis. He is remembered today in the Elbow Valley by the Ings Mine Recreation Area in the Canyon Creek valley

several gas wells were powering homes and businesses in Medicine Hat. And in the Elbow valley, the search for coal was on, led by Dr. George Ings. Within a decade he had a viable mine operating on the slopes of Moose Mountain.

By this time, excitement around oil and gas in southern Alberta was building. By 1912, a gas pipeline had been completed from Bow Island and Medicine Hat to Calgary. When the first oil well in this area was drilled in 1913 in Bragg Creek, Ings Mine was able to supply coal for fuel. Transporting the coal was somewhat easier by this time, since a modest track, mostly high and dry, led up the Elbow Valley from Bragg Creek to serve the Elbow Ranger Station and beyond, and soon a bridge was constructed over the Elbow River (at the site of the current Allen Bill bridge on Highway 66). A local rancher, Jake Fullerton, was hired to haul Ings' coal by wagon into Bragg Creek, and the mine continued operations for another few years. Thus, with its coal supporting the initial development of oil and gas in the area, investigations into the booty that lay within resource-rich Moose Mountain had begun.

As the 1920s rolled around, interest in Alberta's oil and gas again began to expand. In 1927, drilling commenced within the Forest Reserve on Moose Mountain when Moose Dome Oils sunk the first well in the Canyon Creek valley. A second followed in 1936. Meanwhile, Herron Petroleum drilled wells farther east at Station Flats and at what was later called Paddy's Flat. Elbow Oil (supported by the Philip Morris Tobacco Company, no less) sunk a well just south of Bragg Creek, using wood as fuel and water from the Elbow.

Many of these wells were short-lived. The Paddy's Flat well operated for a year only; a teamster, Patrick (Paddy) McCarthy, was hired as caretaker for the well until it should be reactivated. Paddy faithfully remained onsite for 12 years, but the well was never reopened and Paddy finally returned to Winnipeg. The flat, now the site of a large campground, was named in Paddy's honour in 1984.

Cross-section of a river valley showing terraces

As we drive down the broad Elbow valley toward Paddy's Flat and Moose Mountain, the cross-valley profile, with its various bumps and benches above the river, appears clearly. When the massive glaciers receded from this area into the mountains, they left behind thick layers of sand, gravel and rocks over which the early Elbow River ran. At that time, about 15,000 years ago, the river flowed at a level similar to that of Highway 66 today — nearly 50 metres higher and at times about four or five times wider than the present river. Over the years, during times

The Riches of the Moose

What has given Moose Mountain its unique riches? Geologically, this area is termed an inlier — an arched anticlinal structure of Paleozoic rock surrounded by much younger Mesozoic rock as a result of several thrust faults stacking older rocks on top of younger ones. Thus, the famous Rundle limestones (of Paleozoic Mississippian age and mainly found in the Rocky Mountains to the west) outcrop extensively on this dome, which resembles a rocky calcareous island in a heaving sea of green-clad foothills sandstones and shales.

The elevated "Moose Dome" (as it is called in oil and gas circles), an elongated domal structure about 16 by 4 kilometres, has complex geological layering below the surface due to the faulting and fracturing. But the subsurface layers contain a large oil and gas condensate pool. After years of drilling and analysis, the location of this pool — actually four pools — is well known, and Moose Mountain is recognized as a prominent producing field in the foothills. Its coal-bearing rock, discovered by George Ings so many years ago, is no longer mined.

The Too-Popular Canyon Creek Ice Cave

The popularity of the Canyon Creek Ice Cave began when it was visited by a local settler, Stan Fullerton, in 1905, and with its year-round accessibility, it has been popular ever since. The cave was formed by groundwater seeping through the soluble limestone rock of Moose Mountain, enlarging fractures through the freeze-thaw cycle. Its first canyon passage is 150 metres long, with another 600 metres of passage beyond, first accessed in 1968.

Unfortunately, large numbers of visitors in the past led to erosion and destruction of the fragile vegetation on the steep scree slopes surrounding the entrance, making even the cave access dangerous. Now, however, with the road at the bottom of the valley gated year-round, cave traffic has been sensibly restricted to those on foot or bicycle.

---

of warmer climates, high flows in the rejuvenated river cut down into its gravel bed, creating a new river level and a new floodplain. The previous river floodplain then became a terrace of fluvially deposited materials (alluvium), suspended like a platform or bench above and beside the river. Today, a series of identifiable terraces is visible above the Elbow in this part of its valley, and more may be created in future. Some of these terraces are particularly evident in Paddy's Flat Campground where each campsite loop is set on a small terrace, with a short steep grade leading down to the next terrace loop, and finally to the river at the bottom of the valley.

At last, here is the turnoff from Highway 66 onto the Moose Mountain Road. The original access route to the mountain was not here, but was a packers trail leading from the Elbow Ranger Station, up Ranger Creek, and then west to the southeast slopes of the mountain. The current route, west of the original trail, is a twisting road which began as a fire road built in the 1950s. This was improved in the 1970s, and widened a decade later by Shell Oil to facilitate access to their oil and gas drill sites farther up the mountain. The first seven kilometres of the road are open to the public between May 14 and December 1 each year, providing hikers and bikers with a great headstart on their trip to the summit. We are happy to take advantage of it, and drive slowly up the winding gravel track, leaving clouds of dust hanging in our wake.

Seeing a signed viewpoint over the Canyon Creek and Moose Dome Creek drainage, we pull over for a short informational stop. Three hundred meters below us, we can see the Canyon Creek road in the deep valley. Closed to public vehicular traffic, it leads to well sites and to the Ings Mine Provincial Recreation Area. High above the road, a large dark slot in the rock is the entrance to the Canyon Creek Ice Cave, the most visited cave in the Canadian Rockies, after Banff's Cave and Basin.

Canyon Creek road and well site with new pipeline access right-of-way

Evidence of oil and gas operations is everywhere — well sites, compressor stations, pipeline rights-of-way, power lines — all within the background of the thick foothills forest. Industry traffic on the road and in the sky gives the impression of constant activity in the area. So much for the wilderness. At the end of the driveable road, begins the seven-kilometre hike up to the summit, 670 metres higher than the parking lot.

Most of the walk follows a wide fire road, the first part of which heads disappointingly downward — a rocky, rubbly going — for a 100-metre elevation loss. Eventually, the track turns upward and we pass through dense subalpine forest, in some places so thick that it is obvious that fire has been an element in this landscape. In fact, an information sign describes the forest fires that raged at the beginning of the twentieth century, but which today are managed carefully. At last we are past the forest and emerge onto a ridge of alpine grasses and clumps of stunted trees. These are the meadows of the multiple-award-winning cowboy movie *Brokeback Mountain*, filmed here on Moose Mountain in the summer of 2004. The view is spectacular, and becomes even more so the higher we climb.

Although there are many inviting open areas to sit in, we carry on to the plateau below the summit of the mountain, to hunker down amongst the rocks for lunch, and, as on every summit this week, savour the sweeping views. This rocky spot is always windy, and is frankly not that comfortable for sitting either – those crisp limestone shards are sharp! Its alpine austereness, though, is mitigated not only by the view, but by the tiny grasses, wildflowers and mosses huddling between the lichen-covered rocks. A couple of pikas and a ground squirrel scurry busily behind the rocks and the visitor-built inukshuks. Black bears and cougars also live on this mountain, but fortunately, none are in sight today.

It is no surprise that Moose Mountain has been suggested as a provincially significant area. Its special features make a substantial list: its alpine summit, steep scree slopes, the deeply entrenched canyons of Moose Dome and Canyon creeks, the Canyon Creek Ice Cave, subalpine forests and meadows, geological exposures, patterned ground, and at least five rare plant occurrences. Moose Mountain is worth far more than just its underground wealth!

Leading upward to the west, the zigzagging switchback trail to the fire lookout on the summit is clearly visible, and the lookout is our goal. Several young bikers scoot past us on the rocky track, drop their bikes at the base of the steepening trail, and head for the top on foot. It looks like steep and slippery going, but the two rapidly climb

Meeting the Family

up and up toward the lookout, and eventually disappear behind a ridge. Following them a little more slowly, we work our way along the narrow arête path, once in a while glancing down the treacherous slopes on either side as dislodged scree cascades into the ravines below. This is, in fact, the watershed boundary; this arête is part of the divide between the Elbow watershed on our left and the Jumpingpound watershed on our right. And at the top perches our destination, the fire lookout.

Fire! "Managing" this Force of Nature?

Fire has always been present in the prairie and foothills landscapes, modifying and renewing the vegetative cover in an approximately 250-year cycle since the long-ago retreat of the ice from this area. First Nations people used fire as a positive force — for cooking and warmth, but also to open up camp sites, aid in hunting and provide better forage for wildlife. Throughout early history, most fires were started by lightning, but as settlers moved into western Canada, human-caused fires increased in number, and fire became the enemy. As early as 1832, there were penalties for starting prairie or forest fires.

Right after Canada's Confederation in 1867, a federal Department of the Interior was formed, with responsibility for all of Canada's forests. In 1875, the Council of the North West Territories (of which present-day Alberta was a part) passed an ordinance for the prevention of fires. Fire detection and control became an economic imperative, and by 1887, all men in a district were required to assist in fighting fires, albeit with fire-fighting equipment as basic as barrels of water and flour sacks. ➔

Moose Mountain summit and its fire lookout

→ By the 1920s, more serious attention was paid to **protecting** the forests, first by the federal government and then by the Alberta government, which took over the responsibility in 1930. Aircraft were used for fire patrols, and starting in 1921, fire lookouts were built and a fire-control plan established for Alberta's forests. By 1931, as a further prevention measure, fire guards had been cleared on the nearby Forest Reserve boundary. Today, the Forest Protection Division of Alberta Environment and Sustainable Resource Development has a mandate to find and report all fires larger than one-tenth hectare in area, so operates over 132 lookout sites across the province's forested areas — the most intensive such system on the continent. An observer in a lookout can scan an area of over 5,000 square kilometres (about three-quarters of the size of Banff National Park), and detect a fire 30x30 metres (about half the area of a hockey rink!).

As a result of these fire-control policies, forest compositions have changed and not for the better: they are denser with built up high fuel loads, are less diversified, and are more susceptible to insect infestations and wildfire. Recognizing this, many constituencies have implemented a policy shift to the use of controlled or prescribed burns; these better simulate natural processes while preserving aspects of the forests required for human use.

Of the three main natural forces which can affect these watershed forests — fire, wind and infestations — fire is by far the most powerful. Fire can be used to advantage when controlled, but can otherwise wreak havoc. In 1910, after the first survey created the Forest Reserve in the Elbow watershed, a devastating fire swept through the valley. It had started in the mountains of the Kananaskis Lakes area; fanned by high winds, it spread east through Elbow Pass into the upper watershed and down into the Elbow valley past the Moose Mountain area. At that time, fire detection and fire-fighting methods were primitive, and, unlike today, forest fires tended to burn until they extinguished themselves. Standing below the fire lookout and above the verdant foothills spread around them in the bright summer sunshine, I find it hard to imagine the effects of that fire on this landscape — miles and miles of blackened slopes with stark snags and clumps of charred logs remaining here and there. The impact on the watershed must have been immense. And nine years later, just as the forest was starting to regenerate, a second fire roared through the area, starting just below our position on Moose Mountain, sweeping down the Canyon Creek valley and through to Gooseberry near the Forest Reserve boundary.

Former Parks Canada naturalist Mike Potter has collected detailed information on the fire lookouts of the Canadian Rockies, including that on Moose Mountain. In 1928, construction supplies for the first Moose Mountain fire lookout were packed in by horse, up Moosepacker's Trail from the Elbow Ranger Station, about 12 kilometres to the southeast. The lookout was connected to the Station by a telephone line hung on wooden poles brought up by packhorse and erected with great difficulty in the rocky ground. It remained in place for nearly forty years but was eventually replaced with radios.

Pack horses supplying fire lookout and new telephone line on Moose Mountain, 1928 (courtesy Alberta Forest History Photograph Collection, Forest History Association of Alberta)

The 1950s was a busy decade on the mountain. In 1952, provincial fire fighting policy was changed and a more modern era began; instead of only controlling fires within 16 kilometres of roads and major rivers, all fires would now be fought. On Moose Mountain, a second building had to be constructed to replace the original wooden structure, now over 20 years old. Observers received supplies by packhorse once a month and used a rainbarrel to provide fresh water. The lookout building was heated with a woodstove, using firewood cut farther down the mountain. A wind-charging system was installed to provide additional power, but was dismantled in the 1960s when it proved too difficult to manage efficiently. The third, and current, lookout structure was erected in 1974 just south of the 1952 building, which was dismantled. Environmentally friendly solar panels provided a power source in 1983.

Except for a few topographic obstacles, the Moose Mountain observer can see, and is responsible for, all of the forested area in the Elbow River watershed. Equipped with maps, binoculars, radios and firefinders, observers live in the lookouts for up to 180 days during the fire season (usually May through September). Forty percent of Alberta's wildfires — an average of nearly 1,400 fires annually, covering over 210,000 hectares — are detected by lookout observers. Their observations are supplemented by air patrols, ground patrols, infrared scanning, and observations by industry and the public.

The observer sees hundreds of hikers and bikers come and go over the season. And, at the end of each summer, come bunches of trail runners pounding up the mountain in the annual Moose Mountain Trail Race. These hardy men and women compete in a 29-kilometre, 1000-metre-elevation race from West Bragg Creek Recreation Area to the lower summit of Moose Mountain and back, over logging roads, hiking trails, valleys and mountain slopes — many finishing in less than two and a half hours!

Heading down ourselves, we reach the forested part of the trail below the rocky summit and meadows, and again consider wildfire and its impacts. How would a fire 80 or 90 years ago have impacted the watershed we see today? Forests are dynamic ecosystems, constantly subject to natural disturbances after which they rebuild through succession, in which native species continuously

Foothills Forests: All Things in Succession

In the initial stage of post-fire foothills forest succession, sun-loving species begin to grow. Lodgepole pine seed-cones which burst open in the high temperatures of a forest fire have scattered seeds which can germinate and grow quickly. Aspen rapidly send up prolific suckers from their unharmed root system. Plants which grow well on disturbed soil like grasses, the aptly-named fireweed, and shrubs like wild rose and raspberry push up under the tree seedlings. Now that there is food and shelter, woodpeckers, mice, hares, elk, deer and moose return, followed by predators like coyotes, wolves and cougars. After 40 years, the young forest canopy is about 20m high.

The second, 35-year, stage favours shade-tolerant species like white spruce and shrubs growing up as an understory below the pine-aspen canopy, while mosses and herbs flourish on the ground. Multiple layers of mixed vegetation has improved wildlife habitat and the soil is richer due to nutrients from decomposing logs and plants. This mixed forest continues to grow and diversify.

For the next 75 years, in the third succession stage, white spruce continue to push past the pine into the upper canopy, while the pine and aspen, deprived of the sunlight they require and reaching old age (about 80-100 years for aspen, about 120 years for lodgepole pine), die out and decompose on the forest floor. The understory of shade-loving shrubs, mosses and ferns grows slowly, and wildlife that prefers coniferous-dominated forests, such as moose, lynx, grizzly bears and porcupines, move in. About 150 years following the disturbance, the mature spruce-dominated forest has been reached and will be relatively stable until the next disturbance.

regenerate toward the climax forest community. There are three stages of succession in forests like these; most of the forest in this part of the Elbow watershed is less than a hundred years old, and in the second-to-third stage of succession.

Even-aged regenerated lodgepole pine stand

As we continue down through this mid-succession forest, a little striped chipmunk leaps down from a spruce bough and scurries across the trail to mount the nearest trunk and chatter away madly. What happens after a fire to the watershed's animals, or the fish in the streams, or the birds? The vegetation changes seem obvious, but what about other parts of the forest ecosystem and the water?

Fireweed, an early colonizer after disturbance

Fire can have a positive influence on soils by contributing valuable nutrients through ash and decomposing logs. On the other hand, fire-denuded soil is easily eroded by surface runoff and raindrop impact. Resulting siltation of the watercourses degrades the aquatic habitat, which is further disturbed by temporary or long-term increases in water temperature, and by changes in water chemistry from smoke and ash during the fire and from eroded sediments and debris following the fire. Aquatic invertebrates, amphibians and fish are affected, although entire populations are rarely eliminated.

For animals, the effects of fire can also be both positive and negative. Large, fast-moving mammals like deer and elk fare best in escaping the fire; small burrowing animals can avoid the flames if they are far enough underground. Habitat is disturbed if not eliminated by the fire, but later, many herbivorous animals will benefit from the additional minerals available in new plant growth. As wildfires typically burn in a pattern of a few large disturbed areas and many small undisturbed areas, the result can be a diverse landscape of varied habitats and variable-aged stands of trees — a positive effect for wildlife. Birds such as woodpeckers thrive after a fire, eating insects in snags and dead trees. The black-backed woodpecker, for example, seeks out burned forest habitat for its delicious supply of wood-boring beetles.

The second natural force, wind, has also had an impact on this Elbow watershed.

After the wild winter windstorms a few years ago, we could hardly walk on the trails in the spring. Whole hillsides were strewn with spruce, pine and aspen blowdowns, and many of the fallen trees were the largest and oldest in the stands, presumably weakened by age and nearly ready to fall. On the positive side, stands were then opened up to more light and regeneration of the sun-loving species again, increasing biodiversity. All part of the natural cycle.

Another astonishing forest statistic: up to 42 percent of the annual growth of Canada's forests is regularly lost due to catastrophic insect and disease outbreaks, the third natural force affecting forests. In the Elbow watershed, infestations of spruce budworm and tent caterpillar have come and gone over time, but the mountain pine beetle is the insect causing what may be the largest forest insect blight in North American history. After wiping out significant areas of forest in British Columbia, this little black rice-sized beetle has eaten its way through the mountain passes into Alberta, where it is feasting on extensive stands of lodgepole pine. It is a deadly little beast; once it attacks a pine, the tree will die. By 2009, there was only one reported infestation site in the upper Elbow watershed, but the beetle was expected to continue moving eastward, aided by mild winters. Since 2009, fortunately, provincial surveys have shown that southern Alberta foothills beetle populations appear to be

Tiny Troublemaker: The Pine Beetle

A mountain pine beetle infestation is devastating; it not only spoils the aesthetic value of a pine forest, but also reduces the volume and value of a highly merchantable timber species. Therefore, not only those who enjoy the forest landscape, but also the forestry companies holding Forest Management Agreement (FMA) licenses and the government, which wishes to maintain a viable forest industry, complain. In 2007, the Alberta Government decided to reduce the amount of susceptible pine by 75 percent over the next 20 years. All of the pine forests in the Elbow watershed have a high potential for infestation, so these will be subject to treatment of individual trees (by cutting and burning, or applying insecticides), as well as to block harvesting of infested areas by the forest industry.

slightly decreasing year by year; most Alberta hotspots are north of Grande Prairie.

The forest industry is a major player in this Industrial Development Zone of the Elbow watershed. Spray Lake Sawmills (SLS) of Cochrane holds the Forest Management Agreement (FMA) for all of the Elbow River watershed K-Country forests outside the protected areas and parks. An FMA is a 20-year renewable area-based agreement that gives a forest products company the right to harvest, remove and grow timber in a specified area of Crown land, with responsibility for protection of the watershed, environment and wildlife. The agreement stipulates that the forested area is for multiple uses, including not only sustainable timber harvesting, but also recreation, hunting, oil and gas operations, grazing and trapping.

Well, proponents say, logging is not an issue in the Elbow watershed as the upper watershed is quite pristine. We will just ensure that it doesn't deteriorate! But that is not the issue, reply the opponents; a watershed can be ruined on a local basis, when there is too much clear-cutting in one area.

Various-aged clear-cuts on a slope in the Elbow watershed

SLS's current FMA (2001–21) plans extensive clear-cut logging within the watershed, much of it in heavily used recreational areas, for the stated purposes of reducing fire risk near Bragg Creek and of eliminating lodgepole stands prior to mountain pine beetle infestation. The local community and several environmental organizations have strenuously objected, questioning the efficacy of clear-cut logging (the most efficient method for the forest products company). There are, they point out, other means of renewing the forests, reducing the risk of wildfire near communities, and, more immediately, of slowing the infestation of mountain pine beetle. Selective harvesting (more operationally difficult and often increasing fire risk) and prescribed burns are two other options. A second issue involves the increasingly conflicted multiuse nature of these areas, trying to be all things to all people.

Nevertheless, the FMA plans are in place and logging has accelerated in this area. After a full winter's logging in 2012–13, plans for subsequent years include large clear-cut areas between West Bragg and Station Flats, impacting a large percentage of the hiking and skiing trails in the area and lots of recreationists. The provincial government's draft plans for creation/expansion of an Elbow River Provincial Park in a small part of this area of the watershed have not appeased those in opposition. This accelerated clear-cutting appears to be one of the more significant threats to the viability of the upper Elbow watershed, just as accelerated urban development threatens the lower watershed. Maybe we should all review what global water rights activists Maude Barlow and Tony Clark had to say in their book *Blue Gold*, and give this issue some more thought:

> Forests also play a vital role in protecting and purifying sources of fresh water. They absorb pollutants before they run off into lakes and rivers, and like wetlands, they prevent flooding…When forests are clearcut or depleted in nonsustainable ways, the integrity of local watersheds is threatened or destroyed, but when they are harvested wisely or left in their wild state, they can perform their functions as safety valves for rivers and their watersheds.

But as we contemplate this during the downhill walk, I notice something we hadn't seen on the way up — a sour gas pipeline right-of-way, heading up from the Canyon valley to the west, across this fire road and east toward Bragg Creek. Another part of the Moose Mountain oil and gas story.

Beginning in the mid-1980s, plentiful Moose Mountain gas was shipped by pipeline for processing at the Imperial Oil Quirk Creek Gas Plant near Turner Valley to the south. By the late 1990s, however, Quirk Creek was operating nearly at capacity, so an alternative for faster processing was to pipe gas to Shell Oil's Jumpingpound Gas Plant, about 40 kilometres to the northeast. Completed by Shell in 2005, this sour gas pipeline has about 15 kilometres of right-of-way within the Elbow watershed. Moose Mountain is thus fully encircled by oil and gas operations and infrastructure. It somehow seems unfair but inevitable that after a century of interest, old Moose Mountain has been surrounded and taken prisoner by the energy industry.

When planning this new pipeline and considering further development of the Moose oil and gas field, the impact on grizzly bears was considered. Previous studies had shown that the Moose Mountain and Jumpingpound area is one of two major areas for grizzly "berry-and-after-season" habitat within the Elbow watershed. Shell, Husky and Rigel/Talisman have worked with the Eastern Slopes Grizzly Bear Project in an attempt to minimize the effects of their industry on grizzly food and cover, and on bear movements in spring and summer. As part of their environmental programs, Husky and Shell are also conducting winter mammal tracking surveys, spring songbird surveys, and terrestrial insect surveys over a ten-year period beginning in 2004. The objective is identification of changes in these populations over time and in differing environments, and presumably the causes of the changes. All good intentions but the results are not yet known.

Glad to be almost finished the day's explorations, we walk up the last slope to the parking lot while discussing fire, beetles, and logging and other industrial impacts on our little complicated watershed. There are so many sides to every issue. Then, we head for Bragg Creek and a two-scoop ice cream cone, a treat we have been waiting for all week. What a wonderful six days it has been, exploring the river, the foothills ridges and the great watershed "family."

Chapter 6.

Recreating at the River

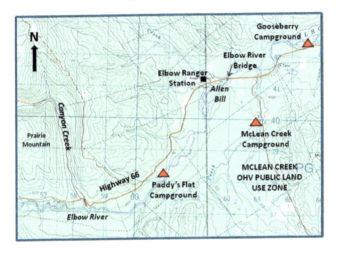

Map of former Elbow Ranger Station and Elbow River recreation areas, including McLean Creek OHV Public Land Use Zone

So many of us enjoy outdoor recreation in the Elbow watershed, whether it be hiking or camping or off-roading or canoeing. Careful as we might be, the impact from our activities can still result in areas within the watershed that are damaged and need the care of volunteers, such as those that assembled on a May afternoon. In the McLean Creek Public Land Use Zone (PLUZ), it was cold, damp and muddy down by the stream and the surrounding spruce-clad hills were shrouded in a light mist. Our motley group of rubber-booted people carried shovels, saws and long lengths of cut willow down the sloppy access road to the creek. Colourful rainjackets

Working for the Watershed: The ERWP

The Elbow River Watershed Partnership (ERWP) was formed in 2002 to bring stakeholders together to "protect and enhance water quality and quantity" in the Elbow watershed, with the vision of providing ample clean water for all. In many watersheds in Canada and elsewhere, water issues have become so critical and potentially divisive that such partnerships are seen as an important approach to achieving collaborative results. Watershed stakeholders are broadly-based, and include four levels of government, private industry (including land developers), individual land owners, environmental and research organizations, universities and First Nations, along with the general public.

Since its formation, the ERWP has built up a significant membership, organized stakeholder and volunteer workshops and meetings, fostered relationships with other similar groups and sponsored awareness-building and educational activities. These activities promote the importance of caretaking of watersheds in general and of the Elbow watershed in particular, where no such effort previously existed. The ERWP also provides input to regional water issues as a member of the Bow River Basin Council (BRBC), the Alberta Stewardship Network, and the Water for Life provincial program.

---

provided the only splash of warmth on that grey day as we started an afternoon's work in the McLean Creek Off-Highway Vehicle (OHV) area. This was the second day of a two-day workshop organized by the Elbow River Watershed Partnership (ERWP) — another in a series of initiatives to improve conditions within the watershed and to raise awareness about water issues.

The previous morning, a dozen other workshop participants gathered at the Elbow Station Fire Base for training in soil bioengineering. In preliminary workshop introductions, the wide range of people participating became obvious. College students, provincial government employees, academics and environmentalists and I all stated a common interest in learning more about landscape restoration and reclamation, and in implementing this knowledge within the Elbow watershed.

Over the day, our instructor reviewed soil bioengineering history, methodology and successful project examples of the use of live plant materials for restoration and reclamation of damaged riparian sites. Engrossing stuff for me and others who believe in the power of things natural and native. The second day's work would involve cutting truckloads of live willow, a fast-rooting and fast-sprouting native species at home in riparian environments, from nearby thickets. The willow was to be used to repair stream banks and adjacent slopes torn up by all-terrain vehicles in the McLean Creek area. And so,

Recreating at the River

the second morning found our group spread out in the muck of a tall willow patch, sawing off the straightest and largest branches possible and dragging them to the road for pickup. Next, we went to the creekside, ready to begin the actual construction.

The ERWP bioengineering workshop took place in the McLean Creek OHV Area, created in 1977 as one of three in Kananaskis Country specified for OHV use. McLean Hill, McLean Creek and McLean Pond are all named for Jack McLean, who established an independent logging operation in the area around the turn of the 20th century. He was the first, but not the last, to float logs down the Elbow River to the up-and-coming town of Calgary. About one-third of this 200-square-kilometre OHV area lies within the Elbow River watershed; the remainder is in the Fish Creek watershed to the south. Within the OHV area is the McLean Creek Provincial Recreation Area which includes 170 heavily-used campsites, a campers' store, two day-use areas for picnickers, and a couple of self-guided walking trails. It is a busy area, winter or summer. So busy, in fact, that its recreational users multiplied by an amazing 450 percent between 1990 and 1999.

After OHVing, fishing is perhaps the next most popular recreational activity here, and receives considerable attention from government agencies. McLean Pond, a small silvery reservoir dammed on McLean Creek in 1983 and stocked every spring by the government with thousands of rainbow trout, provides

Water for Life:
Flowing in the Right Direction

As this new century began, the Government of Alberta faced growing pressure on its limited water supply from a rapidly increasing population, ever-more-serious droughts and industrial growth. Its response was to develop, and in 2003 implement, a strategy entitled *Water for Life: Alberta's Strategy for Sustainability*.

Its three goals — a safe, secure drinking water supply, healthy aquatic ecosystems and reliable water supplies for a sustainable economy — were to be reached through research, partnerships and water conservation. The importance of watershed stewardship groups like the ERWP was formally recognized and supported. In the years since, progress has been made toward all three goals, and an independent international review panel strongly commended the Water for Life collaborative vision and its program as implemented by the Alberta Water Council. They reiterated the 2005 statement of the former premier, Peter Lougheed, that "fresh water is now more valuable than oil," a startling thought for many Albertans at the time.

Now a decade-long program, Water for Life has been reviewed again and is reported to be on-track with most of its goals and key directions. Reviewers praise the philosophy behind it but worry that it lacks the funding to effectively implement its initiatives, relying too heavily on volunteers. There is much work yet to be done.

Fair or Fowl? Alberta's Fish and Wildlife Management History

Conservation and game management were hardly thought about in the late 19th century in Alberta. Overexploitation of game species (like elk, antelope and grouse) was the norm, ignoring the lesson of the 1880s demise of the great bison herds. Early in the 20th century, however, an enlightened few proposed game preserves, such as Banff National Park and the Inglewood Bird Sanctuary, where decimated populations could expand naturally. Fish and game associations also promoted protection and conservation, albeit with hunters and fishermen in mind. Today, Alberta's parks and protected areas cover 12 percent of the land base (more than Canada's 9 percent), providing some needed protection for wildlife. The Elbow watershed has twice that percentage, at nearly 25 percent, mainly in the upper watershed. However, these protected areas are under increasing pressure from development and enthusiastic recreationists, and their wildlife habitat is increasingly discontinuous and fragmented.

Early management philosophies, now considered deleterious, advocated the introduction of non-native species to enhance the local fishing and hunting experience. Global examples of such wrong-headed introductions include rabbits overrunning Australia and zebra mussels clogging the Great Lakes. Perhaps more positively, in 1908 Hungarian partridge from Europe were brought into the Elbow valley, taking pressure off, but not competing with, other game birds like the ruffed grouse. Elk, once extirpated in the Elbow watershed, were successfully re-introduced around 1950, and now have a stable population of over 200. ➔

fishermen with seasonal sport. Upriver on the main Elbow, between Canyon Creek and Elbow Falls, the silt-free water provides valuable spawning habitat for bull trout. Elbow tributaries like Quirk, Silvester and Howard creeks in the McLean Creek area provide critical habitat for fall-spawning bull trout and for spring-spawning westslope cutthroat trout (a threatened species in Alberta). Quirk Creek has been part of a native trout restoration project since 1995, when it was discovered that nearly the whole creek had been colonized by the non-native brook trout. Current thinking says brookies are bad — but why are they even here?

Brook trout, unfortunately, were brought from eastern Canada to the Elbow watershed in 1940, thought to be more competitive than the native bull trout. Competitive indeed, they caused a serious decline in the bull and cutthroat populations. To improve the Elbow watershed fishery, from 1998–2006 volunteer anglers worked with Alberta Fish and Wildlife and Trout Unlimited Canada to fish out the brookies, and were successful in allowing the larger and sportier native bull and westslope cutthroat species to regenerate. Meanwhile, generously stocked McLean Pond diverts some angling pressure from previously overfished trout streams in the area.

Although positive things are happening in the McLean Creek area, not everything is copacetic. Over more than 30 years, the 150 kilometres of designated, but unsigned and unregulated, McLean Creek OHV trails have unofficially multiplied into hundreds more, many of which are causing serious damage to sensitive riparian and aquatic habitats, as well as to upland vegetation and wildlife habitat. Cows and Fish (the Alberta Riparian Habitat Management Society) recently completed an analysis of the riparian health changes between 2007 and 2012 in the McLean Creek, Silvester Creek and Allen Bill areas, as well as in some new sample sites. Despite additional signage and attempted OHV controls, there has been very limited improvement in riparian health. Changes to signage, more public awareness and revised cattle grazing procedures have been recommended. The ERWP has taken on the challenge of signage and public awareness, and will hopefully see early improvement in conditions.

→ Provincial wildlife management continues to change. In the first half of this century, bounties for killing predators (wolves, cougars, coyotes, birds of prey) were promoted, as were protective game regulations like hunting seasons. The bounty policy did not lead to increased game populations, and was discontinued in 1954. By the 1950s, a more conservation-oriented ethic and scientific approach appeared, influenced by famed conservationist Aldo Leopold and others. Game ordinances were introduced to regulate fish and wildlife harvests, evolving into the Alberta Wildlife Act. Despite these changes, wildlife habitat in Alberta continued to decline as agriculture, forestry and petroleum activities multiplied. Today fish and wildlife are managed with a shared stewardship approach. Government, hunters, fishermen, trappers, industry, naturalists and other stakeholders work together to integrate needs and solutions. Emphasis is placed on threatened species of all types, not just game species. This approach is new and not without its pitfalls; its success will be judged over time.

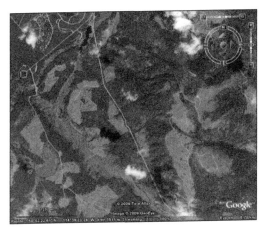

Habitat fragmentation: clear-cuts in McLean Creek PLUZ, with McLean Pond (black) and campground in upper left (Google Earth)

McLean Creek is a busy area. In addition to the increasingly popular OHV use, industries are permitted to operate in the area, and cattle are grazed here seasonally. Husky Energy Inc. and Shell Canada Limited operate pipelines and gas wells, and Spray Lake Sawmills (SLS) cuts timber. All use the existing OHV trails for access, as well as developing their own roads and bridges where required for their operations. Bound by government environmental regulations, and as responsible corporate citizens, these companies strive to protect the areas they are working in, particularly to prevent damage to stream crossings, avoiding them where possible. Such care is vastly different from the time when, for example, sawdust from sawmills was dumped into rivers and streams as a matter of course.

We have learned a lot since those days. Part of SLS's environmental initiative — an intensive aquatic monitoring program in six streams in this area over a ten-year period — studied suspended solids, water chemistry and macroinvertebrates. They concluded that logging did not negatively impact water quality. No surprise there. Meanwhile, a City of Calgary study comparing two Elbow tributary watersheds — unlogged Prairie Creek and busy, clear-cut-logged McLean Creek — shows a distinct difference in surface water quality. McLean Creek has higher levels of phosphorus, nitrogen and total suspended solids, likely as a result of both logging and recreational activity.

Trout Unlimited and others have concluded that OHV traffic has caused significant turbidity in the watercourses downstream of crossings, heavily impacting fish habitat. One major problem in this area in the last few years has been the 5,000-plus off-roaders who gather annually for the May long weekend. Rowdy mud-bogging and the drug- and liquor-related activities of some caused unprecedented ecological destruction, but increased police presence in the past few years has helped to limit the destruction.

To date, such damage has been addressed in piecemeal fashion: some sensitive areas have been fenced off, bridges have been built over damaged stream crossings, and volunteer work parties like this ERWP group have repaired a few of the most immediate areas of concern. Many organizations are recognizing the value of such initiatives. Those partnering with and/or funding the ERWP include Alberta Fish and Wildlife, Alberta Community Development, the Alberta Stewardship Network and Fisheries and Oceans Canada from government. Industrial partners include Spray Lake Sawmills and Husky Energy. Also involved are the Alberta Conservation Association, Friends of the Kananaskis, Trout Unlimited and others. Such partnering is much needed. But what is really needed to prevent ongoing damage and put teeth into environmental protection, though, are strict government regulations regarding OHV use of the area. The government has made a start. By 2012, more stringent regulations were in place, and trails in the McLean Creek area had been classified by vehicle size and re-signed. Enforcement seems to be the outstanding issue.

Riparian area restoration with live willow in McLean Creek OHV zone

Meanwhile, we workshop volunteers got down and dirty, sloshing through the creek, pounding willow stakes, backfilling wattle fences with dirt and stones, and creating pole drainage ditches to steer excess water. Building this infrastructure with natural materials will restore this small destroyed area and its bit of riparian and aquatic habitat. At the end of the long afternoon, mud-caked participants congratulated each other on what we had accomplished. One small step for the watershed!

After changing out of my dirty boots, I drove a kilometre west to sit by what was then Allen Bill Pond to contemplate the implications of the workshop. Is the McLean Creek situation representative of the recreational and industrial pressures within the Elbow watershed? Yes and no. The busiest K-Country part of the watershed (excluding Elbow Falls), extends from the Elbow River Launch near Moose Mountain down to the K-Country border at the Elbow Valley Information Centre. This zone bustles year-round, with forestry and oil and gas industrial activities and with seasonal cattle grazing. Recreationists pursue opportunities for camping, OHVing, RVing, hiking, mountain and road biking, cross-country skiing, snowmobiling, canoeing and kayaking, fishing, picnicking, trail riding, snowshoeing, bird watching, and simply scenic driving. In this ten-kilometre stretch along the Elbow and Highway 66, there are four major campgrounds (including two for groups), two stocked fish ponds, five day-use picnic sites, two equestrian trails, five or more biking trails, seven or more hiking trails, two interpretive trails, many kilometres of OHV trails, the Kananaskis Country Information Centre and RV sani-dump facility and Easter Seals Camp Horizon. In addition, the Elbow Fire Station and helipad, a Department of Transportation facility, four oil/gas wells and a sour gas pipeline are in operation. About 12 kilometres of the Elbow River flows past it all. In summer, the campgrounds are full by mid-week; cars parked by trailheads in hiking or skiing seasons flow out of the parking lots onto the road. How could this not put significant pressure on these beautiful foothills in the watershed?

One reason for this area's popularity is that this is the rich boreal foothills ecological zone, which covers about 20 percent of the watershed and is a transition between the subalpine zones to the west, and the prairies to the east. Rolling hills and ridges, up to 1,900 m high, are heavily treed with lodgepole pine, white spruce and fir on the shaded, moister northern slopes, and covered with mixed aspen/pine woods on their more exposed southern slopes. Even paper birch makes an appearance at lower elevations. There are extensive grassy meadows, both on the flats and on the sunniest, driest slopes and rocky summits.

In the broad Elbow River valley here, a series of terraces supports shrubs (especially willow and wolf willow), aspen and balsam poplar, spruce and pine. Shrub undergrowth is much more varied than in the subalpine; it includes bracted honeysuckle, Canada buffaloberry and the long-blooming provincial symbol, the sweet-scented prickly rose. Low-growing bearberry, with its tiny pink flowers that become bright red berries in the fall, is a favourite food of bears. Called kinnikinnick by First Nations people, its dried leaves and berries were used as tobacco, as a hide tanning mixture and as a dye.

Wildflowers grow in profusion in the woods and meadows. Hikers always know spring has arrived when we see the early purple prairie crocus poking up through dried grasses on sunny hillslopes, followed by the golden buffalo bean, so named by the First Nations people because its flowering season indicated the best time for hunting buffalo bulls. Other easily recognizable favourites include heart-leaved arnica, star-flowered Solomon's seal, twinflower and drifts of tall magenta fireweed. Hardy yellow mountain avens colonize the gravel bars and margins of streams and the Elbow River, turning into fluffy seed puffs in the fall.

Wildlife found here in the Elbow's boreal foothills includes many of the watershed's approximately 80 black bears, a couple of grizzly bears, 230 moose, elk, 800 each of white-tailed and mule deer and a few wolves, as well as cougars, red squirrels, beavers and coyotes. As I well know from my walks, the woods are home to flocks of boreal

Allen Bill, Pond?

When Highway 66 was paved in the early 1980s, crews excavated gravel for the road bed from the river banks just west of the Elbow Bridge, creating a large pit which subsequently filled with water from the Elbow River. This became a day-use recreation spot, and the diversion from the river was sealed off to create a permanent water feature. Allen Bill was named for the retired managing editor/outdoors columnist at the *Calgary Herald* and avid outdoorsman. As the pond was stocked with fish fry each spring to attract fishermen, it became a very popular picnic spot for city folks — a short drive from Calgary, a friendly dappled water body for fishing, wading or watching the ducks and geese, with picnic tables and fire pits, and a short walking trail around its circumference. Even small children have enjoyed examining all the fascinating things to be found growing, flying, crawling or swimming around or in the pond and the river. In winter, families flocked to the area for skating parties and wiener roasts. Now, for the second time there is no pond at all.

In 2005, June rainfall set an all-time deluge record and the river went into major flood mode. The pond and most of the day-use parking lot and facilities were wiped out, and the area closed for repairs. Three years later, the rebuilt and restocked pond and refurbished facilities were back in business, and were as busy as ever — until June 2013, when an epic flood once again removed the pond, the picnic sites and even much of the adjacent highway bridge. Once again, this fishing and picnicking site has been significantly altered, and its future as a rebuilt pond is manifestly uncertain.

chickadees, pine grosbeaks, ruby-crowned kinglets, purple finches, and ruffed, blue and spruce grouse.

In the riparian area and on the river itself, plentiful waterfowl feed, breed, nest, swim and raise their young during the spring, summer and fall. One of the most interesting to be found on the Elbow is the squeaky-voiced harlequin duck, a sea duck considered endangered in eastern Canada and sensitive in Alberta. The blue-black male is easily recognized, with its white neck and face stripes and chestnut/red patches; its colourful plumage resembles its namesake, the brightly-costumed character in commedia dell'arte. The harlequin winters on the warmer west coast, but prefers its Alberta habitat to be cold, clear, fast-flowing mountain rivers and lake inlets, with nearby secluded riparian vegetation for nesting. The little ducks dive for invertebrates on the river bottom, and parents and young alike bounce crazily through the tumultuous river rapids. The Elbow provides perfect habitat conditions in this and its upper reaches. Nevertheless, disturbances by fly fishing, water craft, sediment deposition and riparian cover removal or trampling significantly reduce their available habitat.

This rolling boreal foothills zone is also transitional in terms of climate. Due to its rough topography and moderate elevations, average summer temperatures are typically a couple of degrees warmer than those in the Front Ranges, and 2-4C degrees cooler than

Calgary. In winter, the differences are slightly less. Moisture conditions in this zone are some of the most favourable in Alberta for the growth of merchantable forest species (especially white spruce), thus attracting timber extraction interest. The moderate summer temperatures make hiking, biking and camping inviting. As this zone receives nearly twice as much snow as Calgary and the plains, and only slightly less than the high mountains, it provides excellent opportunities for cross-country skiing, snowmobiling and other snow-related winter sports. Chinook winds during the winter months increase the opportunities for winter activities in pleasant outdoor temperatures. Thus, the combination of agreeable year-round climate, beautiful scenery, proximity to a large urban population and easy access have created the perfect storm for recreational pressure in this boreal foothills part of the Elbow watershed.

Reconstructed and restocked Allen Bill Pond in 2010, looking west

One summer, before heading off on a canoe trip to northern Alberta, our group used Allen Bill pond to do some training. We brought canoes and safety gear one weekday evening to show the canoe novices what we needed to know about paddling, self-rescue, and bailing. As we were paddling about on this benign little pond, a trio of bright red canoes shot past us on the main river, invoking obvious skill to negotiate rocks, shallows and other obstructions in the Elbow. They had likely come from River Cove just upstream, a group campground and one of the main put-in spots for canoes and kayaks.

The reach between River Cove/Allen Bill downstream to Bragg Creek is an intermediate Class II run, with easy rapids and wide clear channels with few sweepers, thus popular with families and paddlers with some experience. Another intermediate reach is found farther upstream, from Cobble Flats to Elbow Falls. At Canyon Creek, just below Elbow Falls, is the Elbow River Launch. From there to Paddy's Flat is about four kilometres of Class III Advanced paddling, including many powerful rapids, exposed rocks with strong eddies, a Class III ledge, and large waves — an exciting run, and not for the novice! All of these reaches are popular with canoeists and kayakers, particularly from late June through August when the water levels are relatively high.

Male mule deer assessing the situation (photo courtesy of Heather Kerr)

Despite the recreational and industrial activities in this zone, wildlife seem abundant enough to be regularly seen by visitors. I spotted two mule deer grazing across the river, apparently pleased to find a few green shoots among the soggy grey grasses from last summer. This boreal foothills zone provides some of the best winter range for the over 1500 ungulates living in or near this part of the watershed. For deer, elk and moose, east-west river valleys like the Elbow offer prime winter range. From this Allen Bill area most of the way up into the subalpine, the Elbow valley provides browse that is close to the secure cover provided by forests and topographic variation, and some protection from cold winds. Its broad south-facing valley slopes have lower snow accumulation and offer warmer conditions for resting.

South-facing slope in the Elbow valley, providing open, warmer winter range

Nevertheless, while the bears are sleeping the winter away, ungulates do experience serious hardships in late fall and over the long cold and snowy winter. Food resources are scarce and often of poor quality, and the environmental conditions are harsh — icy temperatures, biting winds and deep or crusted snow. At these times of year, the animals attempt to cut back on energy expenditures and use their stored body fat to supplement the meagre forage. They look

Can you tell a Muley from a White-tail?

Western mule deer, recognized by their large mule-like ears and a thin brown black-tipped tail centred on a white rump patch, are related to the white-tailed deer (the oldest deer species in the western hemisphere), but have a finely-tuned survival system that is distinctly different from that of the white-tail. They stott! When white-tailed deer are alarmed, they rapidly sprint away, usually downslope, and never look back. When mule deer sense danger, they escape with high, stiff-legged leaps, a fleeing mechanism called stotting, and in an uphill direction if possible. At a safe distance, they stop and look back, those giant ears erect, their curiosity showing. Although stotting requires 13 times more energy than simply running, it has important advantages. The deer gain height to jump over obstacles, see a way ahead, space their tracks more widely (harder to follow) and make rapid changes in direction. Mule deer will also group together to attack predators, something white-tails will not do.

Mule deer are calm and composed, and so polite with each other than they will not look each other directly in the face; their gaze is always averted. They are curious and very tame when not hunted. Males considerately depart the territory of females with fawns, leaving the quality forage for the nursing mothers in the lowlands where cover provides more security. Born in March, fawns lie motionless in hiding for most of their first summer, giving off very little scent to attract predators, until they are large and strong enough to keep up with their mothers. And the mule deer does are exceptional mothers, even responding to distress calls of white-tailed fawns, a behaviour not found reciprocally in the white-tailed does.

for familiar habitats with the best available forage and thermal and security protection, and spend much of their time resting to conserve energy. Often, deer and elk "yard up," forming into groups to help with development of trails through deep snow or to optimize their defence against predators. Long-legged moose have fewer problems moving through snow so can operate more independently.

Good winter range is critical for maintaining ungulate populations, allowing the breeding females to survive and produce healthy calves in the spring. Mule deer produce fawns in wintry March, while white-tailed deer, elk and moose give birth in more salubrious late May or early June, as do the sheep and mountain goats farther up in the alpine areas. Within the Elbow watershed, the mule deer population is increasing annually despite reduction of habitat and increases in their mechanical predator, the motor vehicle.

Mule deer are found nearly everywhere in the watershed — in aspen parkland, foothills and subalpine areas, in higher elevations in summer and lower down in winter. When the lush summer aspen, willow and other tasty vegetation have dried up, mule deer feed happily on Russian thistle, stinging nettles, fireweed, and especially the dry yellow leaves of the balsam poplars which they eat like potato chips! In order not to negatively impact this winter range and disturb or otherwise stress the animals, restrictions have been placed on both recreational and industrial activities in this area of the watershed.

Highway 66 above Elbow Falls is closed from December through mid-June, McLean Creek Trail is closed from December 1 to May 14, and snowmobiling on the Elbow Loop ends on March 31. As forestry and oil and gas activities can provide additional permanent or temporary access to wildlife areas, disturbance to the wildlife, and degradation of the habitat, companies are required to provide industrial access management plans which minimize such disturbances. Winter is the most critical time for ungulates, so industrial restrictions in the Elbow watershed generally apply between January 1 and April 30.

This timing also helps to minimize conflicts with cattle in the Forest Reserve. Since the Forest Reserve was created, eight thousand square kilometres of grazing allotments in Alberta's foothills have provided about 12,000 cattle with summer pasture each year. In the Elbow watershed, cattle are trucked in around mid-May and enjoy the lush bottomland and meadow grasses of the Reserve until October. Hikers have found them and their obvious excretory evidence from the K-Country border all the way upvalley to Elbow Falls for over 100 years, so have to watch their step in the lower meadows and on grassy slopes.

The Elbow Ranger Station and barn under construction, 1915 (courtesy Alberta Forest History Photograph Collection, Forest History Association Alberta)

Across the highway from Allen Bill is Ranger Creek Road, a gravel track leading north to a series of buildings and an equipment yard. An old photograph of the original Elbow Ranger Station in 1915 shows that it was located about halfway down this road, just across Ranger Creek. After the original Forest Reserve boundary was surveyed in 1909, the Elbow area was patrolled by rangers from

The Elbow's Indomitable Ranger Ted

Born in Pakistan, Ted Howard arrived in Alberta in 1901 and married Mary Wilson, a matron at the High River Hospital. After homesteading in the Blackie area for 15 years, and serving in World War I, he joined the Forest Service. The duties of the Forest Rangers were many and varied, including patrolling the area, building trails, fire prevention and fire fighting, and sometimes acting as fish and game control officers – and Ranger Ted was certainly up to the tasks. Ted had an eye for fine horses and was well-respected for his kindness, fairness and honesty; both he and Mary helped local people in need whenever they could. He was a great friend of the Stoney people. Once Ted made a tiny casket out of a packing crate for a Stoney baby who had died in a nearby camp. And when Tom Powderface showed up at the station one bitterly cold winter night, needing assistance with the birth of one of his children in their camp a few kilometres away, Mary bundled up in furs and set off to help deliver the baby.

Ted served as Ranger for nearly 20 years, until 1938. It is a fitting tribute to Ranger Ted's contributions in the early days of the Forest Service that Mount Howard and Howard Creek, both in the Elbow watershed, carry his name. Ranger Creek and Ranger Ridge further commemorate the history of the rangers in this area.

Fish Creek and Jumpingpound. In 1915, though, the Elbow District of the new Bow River Forest, part of the extensive Rocky Mountains Forest Reserve, came into its own. A modest Ranger Station was built on the site of an old winter cow camp belonging to the Fullertons (a prominent family in the Bragg Creek area) and the first Elbow District Ranger installed. After the basic Station house/office was completed, a barn for the horses, a machine and tool shed, a fenced yard and a serviceable bridge over the creek in front of the house quickly followed. By then Ranger Terrence (Ted) Howard had moved in with his wife Mary.

The Elbow River bridge completed by the Forest Service in 1917 (courtesy Alberta Forest History Photograph Collection, Forest History Association of Alberta)

At the time the Station was built, the Elbow District was a remote place. The road from Bragg Creek was little more than a track and the Elbow River had to be forded since there was no bridge in this location. Construction of a permanent bridge over the Elbow became a large, on-going project for the Forest Service and local settlers were hired to help. The bridge was to be a significant structure, made of logs from the area and supported by three giant cribs in the river bottom. Part of the engineering challenge was the development of an access road for the bridge; this required extensive cut-and-fill work down and around a relatively steep hill on the east bank. After this first bridge was completed in 1917, groundwater seeping out of the hill above the access road created impassable ice flows in winter. On one occasion, Ranger Ted and his wagon team slid right off the icy road, over the precipitous edge and down onto the frozen river. Of course, the intrepid Ranger was not daunted; he photographed the ice flow for the record, then managed to obtain funds to construct a culvert to allow the water to pass under the road, effectively solving the problem. Just when the access and bridge were deemed to be in good working order, a major spring flood in 1923 damaged the bridge, requiring extensive repairs before it was again serviceable. By 1928, the bridge had been re-floored and seemed invincible. Until the next disastrous flood.

An Elbow valley road, 1925, with sign saying, "Danger, Go Slow" (courtesy Alberta Forest History Photograph Collection, Forest History Association of Alberta)

Meanwhile, a rustic wooden gateway and twig sign above formalized the entrance to the Bow Forest Reserve. Other signposts denoted Reserve boundaries, and warned of forest fires and dangerous trails. Trail building and then road improvements constituted a significant part of the effort expended within the Forest Reserve.

Who's Running This Show?

In 1930, in a significant step, the federal government passed responsibility for management of Canada's resources to the provinces. The Resources Transfer Act transferred all Crown lands, forest resources, mines, minerals and related royalties, fisheries and wildlife from federal to provincial ownership. Alberta took on these resource responsibilities at the beginning of the difficult depression years, and struggled to manage them efficiently. Through many administrative changes over the next three decades, the provincial agencies for forestry, fish and wildlife have had to work closely together to manage these resources.

In 1969, the Elbow and Jumpingpound forest districts were merged and the new District managed from the Elbow Station. But today, we have come full circle: local management has again moved out of the watershed with the 2002 closing of the Ranger Station (latterly called the Elbow Administrative Complex), including administrative offices for East Kananaskis Country and Alberta Fish and Wildlife. Now the Kananaskis Improvement District (KID), established in 1996 as an unincorporated municipality whose land is owned and operated by the Government of Alberta, provides government and municipal services to K-Country residents. The 50 campgrounds in K-Country (including the six in the Elbow Valley) are currently operated by a contracted company, Kananaskis Country Campgrounds, which has its Elbow offices and equipment on Ranger Creek Road where the old Ranger Station used to be.

The Ranger's work was never done: he was required to prevent or solve all problems, often with the assistance of rangers from adjacent districts. In 1919 wildfires burned extensive areas near Canyon Creek to the west, and at Gooseberry, east of the Station. It was the Ranger's job to detect these fires, and to organize fighting them with the minimal manpower and equipment available — no easy task. By 1924 a new provincial road, the Elbow Valley Trail (the precursor to Highway 66) had improved access up the valley west of the Ranger Station. The 1928 repairs to the Elbow Bridge were completed just in time to begin construction of the Moose Mountain Fire Lookout. By then, the Ranger Station had a telephone, and a line from the Station was run up the mountain for communication with the lookout. There was no shortage of projects in the Elbow District.

Looking down the little road at the spot where the old Ranger Station stood, I wonder what Ranger Ted would think of his Elbow Valley today. I can imagine him, Stetson-hatted, astride one of his fine hard-working horses, looking at today's stream of cars crossing the river on the massive concrete bridge, hurrying up-valley on the wide all-weather highway. Across the river, he would clearly hear all-terrain vehicles buzzing up and down the muddy trails like a swarm of angry bees. He would see helicopters droning back and forth overhead and parking lots full of big SUVs topped with gear boxes. All that activity would have been simply astonishing, and perhaps a little disappointing.

Ranger Ted would have been quite familiar with industrial activities of the Elbow District — oil and gas drilling and timber cutting and even coal mining — but the scale of these operations has expanded exponentially. He would be pleased to see a third-generation fire lookout still standing guard atop Moose Mountain, yet the sight of cyclists, hikers and trail runners racing up and down the mountain slopes would have seemed incongruous compared to his steadily plodding pack trains of telephone poles and construction materials. I suspect he would feel that all of this could not be positive.

Recreational pressure in the foothills of the watershed is a dominant concern today. Reorganization of facilities is currently underway, with the objective of protecting the upper watershed while improving tourist opportunities in the Elbow valley. K-Country has celebrated its 30th birthday, and may now be entering its most challenging years. Its Elbow Valley portion attracts half a million visitors annually, 80 percent of whom are day-users, and Highway 66 traffic has increased concomitantly. Several visitor locations in the valley are chronically congested and the effect of this pressure is becoming increasingly obvious to managers and visitors alike.

In his classic of American conservation literature, *Sand County Almanac*, Aldo Leopold warned about the pitfalls of blasting recreational access into wilderness, worries that hold true today, 60 years later:

> Barring love and war, few enterprises are undertaken with such abandon, or by such diverse individuals, or with so paradoxical a mixture of appetite and altruism, as that group of avocations known as outdoor recreation. It is, by common consent, a good thing for people to get back to nature. But wherein lies the goodness, and what can be done to encourage its pursuit?

How to deal with this pressure in the small and still beautiful Elbow watershed? The answers are not simple, nor even agreed upon. In the 1960s, several small areas along the Elbow River were designated as Forest Recreation Sites; then, during a capital development phase in the 1980s, many were redesignated as Provincial Recreation Areas (PRAs). Alberta Parks manages these areas, with day-to-day operations carried out by contractors, to provide outdoor recreation opportunities for visitors. A total of 616 campsites and 553 day-use units in the Elbow valley do not seem sufficient during our short summer season. Early in the PRAs' development, interpretive signs, evening instructional programs and guided hikes enhanced recreational opportunities; unfortunately program cutbacks of the 1990s eliminated such activities. Today, a few faded signs and empty pamphlet boxes are all that remain and campers are on their own. Maybe we'll all soon be downloading an Elbow Interpretive App? If the educational component reappears, it may be due to the efforts of the Elbow River Watershed Partnership and other such stakeholders.

Proposed changes in the Elbow valley include combining eight small PRAs (Little Elbow, Cobble Flats, Elbow Falls, Elbow River Launch, Ings Mine, Elbow River (Paddy's Flat, Station Flats, Allen Bill), Gooseberry, West Bragg) into a non-contiguous Elbow River Provincial Park. Addition of more trailhead parking and even construction of hotel-type tourist accommodation at Elbow Falls have been mentioned. The latter would be a first in K-Country in the Elbow watershed — the thin edge of the wedge? — but thankfully is not included in the 2013 draft report. The proposed changes would purportedly provide access to more government funds for interpretive programming and facility upgrade and expansion. The result? More visitors accommodated, educated, and happy with their experience? Maybe. Is that part of Leopold's "goodness"?

Change is coming, good or bad: another watershed for the watershed. Somewhat sobered, I climb back into the car and head out of Kananaskis Country, down the smooth wide highway to Calgary, still pleased with the small volunteer contribution that has been made to the health of the watershed.

Chapter 7.

Historic Hamlet on the Elbow

Map of the hamlet of Bragg Creek and surrounding area

On a crisp January morning we are headed west toward Bragg Creek and beyond to West Bragg, anticipating a vigorous outing on the snowy cross-country trails. Recreational opportunities for Bragg Creek residents and visitors alike abound here. Cross-country ski trails are everywhere, many voluntarily groomed and maintained by the local Trails Association in the West Bragg area. Golfing, canoeing and kayaking, hiking, snowmobiling, picnicking in nearby Bragg Creek Provincial Park, fishing, tennis, cycling, and so on — all available on their doorstep. The hamlet lies on the Cowboy Trail, a winding drive through the rolling Alberta foothills on Highway

Stalwart Sam Livingston: From Trading Post to Innovation to Politics

In 1873, 11 years before the Bragg boys showed up, Irish-born Samuel Henry Harkwood Livingston, one of the first European settlers in Alberta, arrived in the area via the California 1849 gold rush, the Cariboo and Fort Edmonton. This tall bearded adventurer, entrepreneur and innovator came south with his wife and the first of their 14 children to establish a homestead on the Elbow River 12 kilometres north of present-day Bragg Creek. He traded buffalo skins and other goods with the plains Indians (who called him Big White Devil) beside the Roman Catholic mission, Our Lady of Peace. Soon, however, new opportunities beckoned and the family moved downstream.

Sam Livingston, doughty Elbow valley pioneer (Glenbow Archives NA-94-1)

In the broad grassy Elbow valley, on the site of the present Glenmore Reservoir, Sam set up a trading post to provide the newly arrived North West Mounted Police at Fort Calgary with vegetables, milk and meat. ➔

22, featuring unparalleled vistas and many options for recreation related to the area's rich ranching history. And this scenic part of the watershed has been the setting for numerous television shows and movies: *North of 60*, *Wild America*, *Storm*, and of course, farther up in the watershed, *Brokeback Mountain*. It is a special place, and is still infused with pioneer spirit.

Bragg Creek, an unincorporated hamlet administered by Rocky View County, has a history that belies its size. Pioneer settlers first came to this area over a century ago in 1894, as Calgary was maturing into a city. Two enterprising teenaged boys, Albert Warren Bragg and his brother John (Jack), arrived from Nova Scotia, having run away from their new stepmother to seek adventure in Canada's West. The Elbow River watershed, its grassy meadows often cleared of trees by wildfires, seemed an attractive setting for ranching. They applied for a homestead near a small stream in this West Bragg area, and briefly set up housekeeping in what would be the centre of settlement for the next 20 years. When the Dominion of Canada surveyor, A.O. Wheeler, came through the area that same year, he noted their homestead and gave the unpretentious little creek running down from Moose Mountain into the Elbow the name of Bragg Creek to recognize these early settlers. The young Bragg boys, however, found ranching more challenging than anticipated. Winters were long and harsh, snows deep and lasting long into the spring, and grasses for livestock feed seriously limited due to the short growing season. Finally Warren and Jack grew so homesick that they left the area for their East Coast home.

Before 1900, few sections had been homesteaded in the area of the Elbow River and Bragg Creek confluence. Sandwiched between the Tsuu T'ina Reserve to the east, and the densely forested rolling foothills to the west, it was a difficult area in which to establish traditional agriculture on a scale to support a family. Although by this time the Springbank and Elbow Valley areas farther east in the watershed had growing populations and an established agricultural base, the Bragg Creek area had a harsher climate and access to and from the major urban centre of Calgary was more challenging. Despite these difficulties, local ranching and farming began to grow after 1900, and a small community formed. Life in this area was rugged and isolated. Clearing land in treed and/or boggy areas to establish a farm or ranch was arduous at best. Still, its advantages were abundant fresh water, large trees for building homes and barns, and trout, ducks, grouse, deer and moose as a ready food supply.

Contact with the outside world was a struggle. Local residents had to venture several hours, usually on horseback, to the small post office at Jumpingpound (about 15 kilometres north) over rough trails, river crossings and swamps to get mail and supplies. In 1909, forward-thinking Bill Graham, who had settled near the present Saddle and

➔ On his small dairy farm — the first in the Calgary area — he also grew grain and produced fruit from 350 trees imported from Minnesota, the first farmed fruit in Alberta. Always ahead of his time, he was the first to introduce farm machinery to the area. He was also fiercely active in community affairs as a founder of the Calgary District Agricultural Society, the Alberta Settlers Rights Association, the Calgary branch of the North-West Territories Stock Association, and the Glenmore School (mainly populated by his own children), and as a delegate to the Conservative Party convention in 1896. He finally achieved prospecting success with a gold strike on the South Saskatchewan River.

Sam died suddenly in 1897 at the age of 66, not long after his fourteenth child was born. Part of his Glenmore home is now located in Calgary's Heritage Park and his gravestone is in their cemetery. A provincial fish hatchery, an office tower complex and a Calgary elementary school are named in his honour, and a large bust of Sam is sited at the Calgary airport. Who in the area does not know something about redoubtable Sam Livingston?

The Livingston house and barn preserved at Calgary's Heritage Park

Sirloin Ranch, set up the first post office in his kitchen and fetched mail for local residents from Jumpingpound once a week. While agriculture was slowly being established, exploratory coal mining and oil and gas development came and went, providing the local settlers with some supplemental income from part-time employment. To the west, the establishment of the Bow River Forest Reserve in 1909 and the Elbow District of the Bow River Forest in 1915 also provided work for the locals in road and bridge building, freight hauling and related jobs. Daily life was becoming a little easier.

Those barely imaginable hardships of homesteading fascinate me today, sweeping past in a warm car and gazing at those same snowy fields pocked with willow, spruce and aspen. Moose Crossing signs and two coyotes tracking north away from the road attest to the wildlife still in this area, even though it is more populously settled today on farms and acreages. Farther along, the road changes to gravel as we enter Kananaskis Country and then the West Bragg Creek Day Use Area. On a sunny winter weekend, the large parking lot is crammed with cars, spilling out along the road for 200 metres. But today, there are but three other cars and the area is quiet except for the squawk of the odd magpie.

Down the ski trail, we remark on the changes that have occurred here over the past few years. The trail we are following has been upgraded from a rough track to a solid well access road, and the rickety log bridge that used to cross the creek at the bottom of the hill has been replaced with a stable concrete structure capable of bearing heavy loads. All this bespeaks the oil and gas and Spray Lake Sawmills development that is ongoing in this part of K-Country. And looking at the foothills slopes to the west, there are more clear-cuts and seismic line rights-of-way in what had been unbroken forest not long ago. Change is inevitable, perhaps, but the most recent government-approved clear-cutting of much of this West Bragg area by SLS, politically justified by the advancing mountain pine beetle and reduction of fire risk, has raised a great to-do.

"Save Kananaskis" sign, part of the Tagatree and Create-a-Park campaigns

Beneath its calm rustic veneer, Bragg Creek harbours a dedicated cell of environmental activists. Exceedingly well-organized, they surface when threatened and bare their collective teeth against invaders, whether these be government or industry or even other well-meaning citizens. Serious issues — water quality, an area structure plan with land use changes, increases in forest clear-cutting, oil and gas development — have impacted their community. These are pounced on, researched, brought to public attention by creative campaigns, advocated, and then often a pro-environment solution negotiated with the attacker. A recent Create-a-Park campaign lobbied to create a provincial park in the currently "unprotected" Moose Mountain-West Bragg Creek portion of Kananaskis Country, a stand supported by many in the region but not by the provincial government to date. A Sustain Kananaskis initiative

"Go-get-'em" Jake and the Fullerton Legacy

From 1913 on, Jake Fullerton seemed to have a finger in every pie in the Bragg Creek area. In addition to his ranch (the Circle 5, now Elkana Ranch), he ran a sawmill, had a teaming business, supplied firewood to settlers in the Jumpingpound and Springbank areas, transported freight for oil wells, built roads and hauled timber for the Elbow River government bridge upstream. In the 1920s, he started a dude business, taking wilderness pack trips up the Elbow valley to Elbow Lake and the Kananaskis Lakes, and hosting rodeos on his ranch. He built a unique dance hall (the Round Hall), an octagonal log structure which still stands today where Bragg Creek meets the Elbow, disused since the 1970s. Jake also built and operated the Upper Elbow store in the Bragg Creek hamlet. A wooden swing bridge across the Elbow connected the store with the dance hall after 1932, when the main bridge in that location was wiped out in that year's big flood. And on the west side of the Elbow, clever Jake sold cottage lots to a new breed of recreationist from Calgary.

Jake was always community-minded. In 1930, he sold 20 acres of his land to Camp Cadicasu, a Roman Catholic Diocese of Calgary youth camp operated on this site until 2009 when it moved north to Jumpingpound. This was just one of several camps for children on the Elbow River in the Bragg Creek area. Kamp Kiwanis was founded in 1951 by the Kiwanis Club of Calgary for disadvantaged children. Easter Seals Camp Horizon was built in 1965 for special needs children. And Camp Gardner, a Scout camp near Pirmez Creek, is still operating on land donated by the Gardner ranching family. ➔

brought hundreds of people out to an Open House on extensive clear-cut logging in the West Bragg trails area, and continues to badger the government for more community input and more appropriate environmental solutions to forestry development in the area. There is no shortage of environmental issues to be addressed.

On the drive back to Bragg Creek, still talking about its history, we stop and walk down to the bridge spanning the ice-covered Elbow. Upstream just beyond the bend, Bragg Creek (the creek) meets the Elbow River. While the earliest homesteading had focussed on this little Bragg Creek much farther to the west, before long the land where it meets the Elbow River became the community's centre.

By the 1880s, settlers downstream in the watershed, on the plains nearer to Fort Calgary, had recognized the potential of these upstream forests and the river flowing through them. Thomas Kerr (T.K.) Fullerton, an early homesteader on the Elbow River near the present Twin Bridges on Highway 8, began logging in the Bragg Creek area, even building a sawmill to process the logs. Like Jack McLean further west, he floated logs down the Elbow into the newly incorporated prairie town of Calgary, where they were sold for construction and firewood. By 1907, T.K. decided he liked the area enough to homestead the quarter-section at the Elbow and Bragg Creek confluence, thus firmly rooting the Fullerton family in this area where they have remained for over a century. T.K.'s fourth son, Ernest (Jake), a former light heavyweight boxing champion of western Canada, bought this quarter from his father in 1913, making it his permanent home and raising his six daughters there. Four other Fullerton brothers settled nearby.

Due to the undeviating, yet often impractical, nature of the western grid system of land survey, the Fullertons' home quarter is bisected by the Elbow. A reliable source of good water is one thing; being physically separated from part of your land by a broad, fast-flowing river is quite another. Right away, Jake set about building a bridge to connect the two parts of his valuable homestead land. That done, he got on with other things, and he was a busy man.

→ After Jake's retirement, Mount Fullerton, high up in the Elbow watershed in the Fisher Range, and the Fullerton Loop hiking trail in Kananaskis Country were named in his honour, testimonials to his invaluable contributions to the development of this region.

Jake Fullerton's Round Hall, seen here in 2009, over 90 years old

"Steady On!": The Unwavering Whites

After arriving from eastern Canada in 1910, Harry and Ida May White managed a small Bankview store in Calgary, and then moved out to the Bragg Creek area in 1915 on the suggestion of one of their customers, Jake Fullerton. Homesteading on the quarter immediately east of the Fullertons', the Whites cleared the land and set themselves up in a tent and board shanty near the Elbow until their log home could be built. By 1918, they had taken over the post office and mail-carrying contract from Bill Graham, including the year-round weekly journey to Jumpingpound and later to Calgary when the route changed. Eventually, they built a separate post office and store, moving the business out of their parlour. Harry's big polished McLaughlin touring car, used to fetch the mail, was a local marvel.

Soon, however, Harry's precarious health deteriorated and he died in 1925. Although Ida May lost the postal contract that year, she carried on with the store and added a petrol pump, providing the only filling station service in the region. Until 1953, the post office contract alternated between Ida May and the Fullertons, depending on which political party was in office. Just as she seemed to be successfully supporting herself, on a hot July day in 1930 Ida's little log house burned to the ground. She was not homeless for long, however, as faithful friends and neighbours built her a new house which she christened *Wake Siah Lodge* (meaning "not far" in the Chinook language). Ida May enjoyed a brief remarriage, and died in 1953.

In 1933, responding to a request from a group of Calgary women led by sisters Catherine and Mary Barclay, Jake set up a 4m by 5m rented tent on his property to serve as the first youth hostel in North America. The tent, along with a horse and a Model T Ford, serviced the beginning of a vast youth hostel movement, allowing young people to travel to rural areas during the Depression and experience the wonders of nature. This hostel movement first expanded to Jumpingpound, Banff, Morley and Canmore, and then to eastern Canada as the Canadian Youth Hostels Association. In 2012, Environment Canada named this Bragg Creek location an historically significant site, stating that "our system of hostels rests on the visionary foundations embodied in the Bragg Creek Youth Hostel." A proud bit of heritage!

From all of these activities, Jake knew the country like the back of his hand. When an Avro Anson training flight from Calgary went missing in 1941 somewhere high in the watershed, Jake was asked to lead the search party. Within two days, he had located the one surviving airman and devised a clever stretcher arrangement to transport him through the rough bush; the fortunate airman recovered in a Calgary hospital. A memorial to the two crew members killed in the crash is found east of Mount McDougall on Powderface Trail near Canyon Creek.

After the Fullertons came Harry and Ida May White, a young couple who had come west to Calgary in 1910 for Harry's health.

The Whites homesteaded in Bragg Creek and operated a store and post office on what is now White Avenue. As we drive slowly down this street, we note that their Wake Siah Lodge is once again home to a family, and that the Whites' old store, after nearly a century, is now a summer-time ice cream and snack shop. These sturdy shelters were built to last!

By the 1920s, a newly graded dirt road from Calgary all the way to Bragg Creek was passable by motorized vehicles at most times of year. South and west of the tiny community, however, only wagon trails existed; cars and trucks had to transfer their loads onto horse- or mule-drawn wagons for further transport. Nevertheless, building blossomed during the 1920s; many long-standing area families arrived during this period.

The Elbow River near the Bragg Creek Trading Post, looking south over the riprap on this curve in 2008

As the population of the local area grew, so did the need for a school. Initially, children were either home-schooled or travelled to Springbank or Elbow Valley, often boarding there in the winter when daily travel was too difficult. The first school was built in 1914 on the West Bragg road; in 1931, a second local school called Two Pines was established south of Bragg Creek, with a teacherage added later. When this school was closed some years later, practical residents brought it to their community centre to serve as a skating shack. But it was not until 1997 that the hamlet had a local modern school (Banded Peak Elementary) for the more than 200 elementary children in the area. Somewhat oddly, though, there has never been a church built in the hamlet, possibly because of the abundance of churches in nearby Springbank.

The Bragg Creek Trading Post, 2008

We park at the Trading Post, in the small area sandwiched between the store and the river, and cross the road to the river edge before entering the store. A man picks his way along beside the iced-up river, preceded by a shaggy golden retriever who is busily sniffing out treasures amongst the snow-covered rocks. In the trees across the river where the old swing bridge used to end, we can glimpse the Round Hall, hunkered down a little sadly in the snow. Bragg Creek lives intimately with its river, and its river has shaped the hamlet in every way. Throughout the development of this community and surrounding area, the Elbow flowed right through the middle of things and could not be ignored. Like rivers running through so many other communities, it has been a mode of transportation for people and goods, a source of recreation, a major barrier which needed to be bridged, a water and food source, and sometimes a pathway into the foothills and mountains to the west.

One of the first issues facing the tiny river-straddling community was a reliable means of crossing the Elbow. There were a few good spots for fording the river by horse or wagon, but at many times of year even those fords were unusable. In 1913, a settler-built bridge was placed across the river about where we are standing, near the current Trading Post, but it was soon washed out. A replacement bridge was similarly lost in the flood of 1916. After 1916, a third bridge — the first built by the government — was constructed, this time with much heavier timbers and cribs filled with rock, and it managed to stay intact until the great flood of 1932, the largest recorded flood to that time.

The winter of 1931–32 in Alberta was long and cold, but not unusually so. The ranchers and farmers in the Bragg Creek area managed to keep cattle fed on their winter

Watersheds and Floods: It Just Depends

The Elbow River watershed sits in a northern climate, on the east slopes of a north-south-trending mountain range. As in other northern watersheds, snow and ice accumulate over the lengthy winter period and river discharge is then at its lowest. In spring and early summer (usually late May to mid-July on the Elbow), warming temperatures release the water in the snow and ice. Snowmelt in the mountains of the upper watershed combined with typically high levels of June rainfall produce the highest flows of the year, often causing flood conditions in the river. But as the summer progresses, snowmelt is finished, precipitation decreases, and discharge lessens; the river drops back to much lower levels. The elements that contribute to discharge — precipitation, groundwater flow, water storage and evaporation — vary widely throughout the year, and from year to year, causing the amount of water in a river to similarly vary.

The Elbow watershed is shaped rather like an unevenly knotted bow tie, with the west side of the bow larger than the east side, and a constriction at the knot in the middle, right at the hamlet of Bragg Creek. The larger area contributing to runoff within the watershed is upstream of Bragg Creek, and spring snowmelt and June precipitation are large contributing factors there. Over the community's history, spring floods have wreaked considerable havoc as they roar through the hamlet.

pastures, and to keep their community life functioning as usual. Upstream on the Elbow, a work camp for 25 unemployed men, set up to help them survive the Depression years, was supervised by Ranger Ted Howard. They spent the seemingly endless winter living in tents and making a fireguard along the Forest Reserve boundary, thinning trees and cutting firewood. By winter's end, they had completed the task, and the thoughtful Ranger had some of the firewood delivered to widow Ida May White.

In April, however, spring still seemed a long way off. On Thursday, April 21, a massive snowstorm dumped a 24-hour record of over 50 cm of snow on the area, choking access to horses and cattle in pastures, newborn calves and hay supplies. Over the next few days residents struggled to return roads to normal, reach stranded cattle, and retrieve supplies and mail. Three weeks later, the snow had gone but the ground was saturated with moisture from twice the normal amount of precipitation since December. Nevertheless, by the end of May, cattle had been moved up to their summer pastures and crops had been planted. Soggy though the ground was, it seemed that spring had truly arrived and the long-anticipated summer was not far off.

After enjoying a pleasant spring morning, on Tuesday afternoon, May 31st, ranchers glanced at a darkening sky and felt the first drops of rain. Through the evening, light rain continued to fall, a normal occurrence at the beginning of June, the rainiest month of the year. By Wednesday morning, however, it appeared that this was not an ordinary June shower. Throughout the day, the rain poured down, reaching a record daily total for June of 79.2 mm. After dealing with initial rain-related concerns like leaky roofs and muddy trails, the attention of the residents rapidly switched to the Elbow River tumbling through their community. As the day progressed, the river level rose rapidly in the driving rain and the sound of the river became deafening. By nightfall, the sound of boulders grinding down the river bed and giant trees crashing into the sturdy government bridge and the river banks could be heard above the roar of the debris-clogged water.

In the most immediate danger were the two community stores and a log home located on the east side of the river. As former residents tell it, the swirling waters climbed to flood the area behind the stores at the same time as it was gouging out the river bank in the front. The Badleys and their baby, occupants of Jake Fullerton's large store and post office, had water reach their front porch before they realized the danger they were in. In the log home, the Harwoods and their daughter were paralyzed with fear as they watched the flood slosh into their kitchen. Fortunately, Shorty Mitchell, a local farmer who had married into the Fullerton family, was returning to his young wife and week-old baby boy on the east trail from Calgary, driving a team and large wagon. Shorty spied a potential disaster, veered his team around to approach from the back through the still-rising water, and hauled first the Badleys and then the Harwoods into the sodden wagon. Plunging once again into the icy flood, the hardy horses pulled the heavy wagon to higher ground and Shorty deposited the two cold, soaking-wet families into the safety of Wake Siah Lodge and the kind ministrations of Ida May White. Throughout that day and the next night, the thunderous waters rose still higher, until with a roar the bridge gave way and both stores and the log home were swept downstream. All of their possessions were lost in the flood; to compound the tragedy, Mrs. Harwood died soon afterward of pneumonia.

Epic Elbow Floods: A Nasty Trend?

The Elbow River watershed is susceptible to extreme flooding. In 2005, the area received the largest amount of total June rainfall in its recorded history: 248 mm of rain from three large rainstorms. Flows on the Elbow were measured at 13 times the normal June levels, and damage caused by the resulting floods was extensive. In Bragg Creek and upstream on the Elbow, the flood forced the evacuation of many residents from their homes, damaged trails, roads and bridges, and took out the berm beside the popular Allen Bill Pond so that it became part of the river once again. Downstream, in the townsite of Redwood Meadows on the Tsuu T'ina Reserve, a large protective berm and nearby mature spruce were ripped out and many homes flooded. Further down the river, for the City of Calgary's operating expenses and infrastructure impacts alone, damage was estimated at $17.2 million; costs for insurance claims and damage to 40,000 homes have added significantly to that. →

Rain continued to fall for the next three days until Saturday, June 4th. Peak flows at that time were a record 836 cubic metres per second (cms), more than triple the peak flows recorded for later floods in 1948 and 1963, and 35 times the average discharge for June at Bragg Creek (about 24 cms). Downstream from Bragg Creek, more flood damage occurred but a major disaster for the small city of Calgary was fortunately avoided. The previous year, construction had begun on an Elbow River dam, to create a reservoir as a water supply for the growing city. At the time of the flood, the dam had been completed but the reservoir area was as yet dry; the empty reservoir behind the dam was thus able to absorb much of the floodwater roaring down the Elbow that June. The 1932 flood discharge level was not exceeded until 2013, 81 years later.

After that devastating experience, the Elbow River bridge location in Bragg Creek was altered. The new steel bridge, anchored in bedrock, was constructed further downstream, at the end of the current Balsam Avenue. This structure was a success, withstanding major floods in 1948 and 1963. It was replaced in 1983 by the current solid concrete structure in the same location, which has survived both the 2005 and 2013 floods. To minimize bank erosion, rip-rap was placed in critical locations in the hamlet, as on the east bank in front of the Trading Post, but virtually all of that is proposed to be replaced and reinforced after the 2013 flood.

→ While that flood did cause significant damage, it was greatly surpassed by events in 2013. On June 20, a Pacific Northwest weather system was stalled over the Front Ranges. In the Elbow watershed, already saturated ground, the remaining snowpack in the mountains and the steep upper watershed slopes exacerbated the effect of 16 hours of heavy rain — a year's worth of precipitation in only two days. The river swelled to unprecedented levels, peaking at 959 cms (triple the 2005 level) at Bragg Creek. Properties in the flood plain were demolished, berms and other barriers breached, Highway 758 eroded in the hamlet. Downstream in Calgary, more than 75,000 residents were evacuated, the downtown and 26 neighbourhoods were flooded, and flood waters poured over the Glenmore Dam in this 1-in-500-year flood. Insured property damage in southern Alberta topped $1.7 billion, and the 2013 floods became the costliest disaster in Canadian history.

With this 2013 call to action, the Alberta government has embarked on flood mitigation measures in the Elbow watershed, considering construction of berms in Bragg Creek hamlet, a dry dam near McLean Creek, a storage reservoir in Springbank, and a diversion tunnel from the Glenmore Reservoir to the Bow River. Time will inevitably show how successful these measures are, should they be implemented.

Flooding not only affects infrastructure, but also has ecological and hydrological impacts on the river and its watershed. Bank erosion, sediment infills, growth of algae and aquatic plants, changes to the bed profile and unhealthy high concentrations of nutrients from sediment all occur. Such impacts are to be expected as part of the natural cycle of rivers. The bed profile and sediment deposition in my downstream section of the Elbow are certainly different after every June freshet. Most of these changes are caused by flooding surface waters within the floodplain of the river. Within the city, however, where such waters are normally more controlled, the majority of damage results from rising groundwater, which floods basements and other structures.

Outside, as we had seen from the bridge, the Elbow is covered with layer upon layer of ice, to the point where its surface is high above the level of the flow of water seen through the few open patches. Given that winter is a low-flow season when there is no ice or snow melt happening within the frozen watershed, where is the water forming the ice layers coming from? The answer is groundwater! Not only are the surface waters of the Elbow River and its tributaries of interest and concern to the residents of the hamlet and surrounding area, but equally so is the water underground. This groundwater is less well understood, but no less important to the health of a watershed and its residents. And the quality and quantity of groundwater

The Venerable Trading Post

I visited the Trading Post on a winter day a few years ago. It was hard to imagine the Elbow running as high as it does in June, pounding and bashing up against the rip-rap bank. For the moment, the Elbow passed peaceably under its layer of ice past the Trading Post on White Avenue. Inside, a wood stove crackled, exuding waves of heat and a pleasant smoky aroma. Interesting goods lay on every shelf, exactly as one would imagine finding in an old trading post. From beautiful beaded deerskin jackets, to boxes of pink ladies bloomers, to intricate native jewellery and moccasins, there was something inviting at every turn.

The flood-devastated Trading Post, seen in July 2013

Formerly the Upper Elbow General Store (opened in 1925), this store has been operated by Jack Elsdon's family continuously since 1932. It was built in a judicious location near a natural ford on the Elbow and on the old Stoney Trail, a north-south route used by the Stoney Indians from Morley. The Stoneys would camp beside the river just upstream in the area of the present provincial park, and trade deerskins and furs for food and other staples like bacon, baking powder and ammunition. Many of the store's goods are still supplied by First Nations in Alberta and elsewhere, making it unique in the area. ➔

→ The bad news is that the Bragg Creek Trading Post — the iconic landmark in Bragg Creek — was literally trashed by the June 2013 flood. At this curve in the Elbow River, the raging water overtopped the rip-rap and the road and submerged the Trading Post and its adjoining residence to the eaves. Outbuildings were swept into the river, trees uprooted, and historic signage and equipment torn away. The resolute owners are rebuilding.

within the Elbow watershed have become environmental and political issues only within the last few years.

The emergence of groundwater as an environmental concern has been relatively recent. Using aerial photographs, geotechnical data, water well logs and ground surveys, the Canadian government has begun a long-overdue national inventory of major aquifers right across Canada. This inventory should be completed between 2024 and 2031, much behind that of the United States which covets Canada's water resources. Considerable hydrogeological inventory and analysis is being carried out on the Paskapoo Aquifer in southern Alberta, the most intensively used aquifer in the province. The provincial government is also reviving its interest in groundwater mapping and analysis, including the Elbow River aquifer.

Diagram of aquifer/river water exchange where the direction and amount of flow varies with the time of year

Because it sits on relatively impervious sandstone, the City of Calgary relies on groundwater found in the valleys of the Bow and Elbow rivers where sand and gravel sediments from ancient and modern rivers have provided relatively broad permeable alluvial aquifers. The shallow, unconfined Elbow

River aquifer, part of the critical system that supplies drinking water for so many Albertans, underlies about five percent of the total area of the Elbow watershed. This alluvial aquifer, recently mapped from Cobble Flats in Kananaskis Country to its confluence with the Bow River in the City of Calgary, is located on both sides of the river, narrow in the upper watershed and up to two kilometres wide downstream. Most of the mapped aquifer is in Rocky View County; the remainder is in Kananaskis Country, with smaller portions in the Tsuu T'ina Nation and the City of Calgary.

Hydraulically connected as they are, contamination of either the Elbow River or its aquifer will automatically affect the other. Studies by water quality specialists at the City of Calgary have found that water quality in the Elbow River upstream from Calgary has deteriorated over the past decade or so; concentrations of phosphorus, turbidity and fecal coliform bacteria have been increasing, although their sources have not been confirmed. Even so, for many years the hamlet of Bragg Creek, sitting prettily atop thin soil over the alluvial aquifer (about two metres below the surface) with no municipal water distribution or treatment system, has been significantly affected. Due primarily to septic contamination of the aquifer from which residents draw well water, a boil-water advisory has been in place since the mid-1970s; for nearly 40 years, residents and businesses have had to truck in potable water from Calgary or

Groundwater: Hidden Treasure

Because it is not easily observable, water which lies below ground commands much less attention than the lakes and ponds, rivers and streams that are visible on the surface. Groundwater lies in the pores and fractures of gravel, sand and rock below ground level, and is a critically important source of water for agriculture, industrial uses and drinking water in many areas of the world. In the United States, 21 percent of all water used comes from groundwater sources, and nearly all of that from freshwater aquifers. Groundwater supplies about one-third of Canada's population with drinking water, and is the primary source for irrigation and livestock watering. In Alberta, only three percent of water needs are supplied by groundwater at present; nearly half of this is for industrial uses like oil recovery through injection, and an additional one-third is for agriculture. In future, however, a predicted continuing decrease in the Rocky Mountain snow and ice supplying most Alberta rivers, and growing competition for surface water licenses, will make the province increasingly dependent on its groundwater supply.

Groundwater becomes surface water when the upper limit of the aquifer reaches the surface, as in lakes, streams and wetlands. This discharge into surface water drainage systems provides water flow between rainfalls and snowmelt, when these drainages might otherwise be fully or nearly dry. Water discharge in the reverse direction, from surface waters into groundwater, also occurs at times of high river flow, as during the sudden spring/early summer freshet. At these times, the river recharges the aquifer. →

→ Understanding this hydraulic interconnection between surface and ground water systems is critical for management of the water resource in any watershed. Groundwater is highly susceptible to contamination from a host of surface water-related sources: fertilizers, pesticides, oil and other industrial spills and leakages, road salts, intensive livestock operations and septic fields. Cleaning up a contaminated aquifer is difficult and costly. The quantity of groundwater is not only affected by natural processes such as climate, but also by human activities, including urbanization (impervious surfaces like pavement restricting the recharge of groundwater by precipitation) and over-extraction by industries or municipalities.

Cochrane. The plume of contamination also flows downstream from the hamlet, calling to mind the old environmentalists' motto, "We all live downstream." Water quality in the river has been affected through its interconnection with the contaminated groundwater. Natural disturbances like bank erosion during flooding, and human-caused disturbances such as removal of forest vegetation or road development also impact water quality.

Land use within the Elbow's floodplain has been tagged as one of the most critical sources of contamination, even though specific point sources have not yet been identified. Once again, such research points out how the elements of a watershed are interdependent, connected as they are through the flow of water. Recommendations in the Elbow River Watershed Management Plan (and approved by Rocky View County Council) call for restriction of most, if not all, future development on top of the Elbow's aquifer, a good forward step. Meanwhile, for Bragg Creek, while the contamination problem apparently cannot be addressed at present, the symptom is being dealt with. At long last, a water treatment plant has been constructed to provide a local source of potable water, with eventual piping to individual residences and businesses. In addition, a wastewater treatment plant will treat wastewater from residences and businesses prior to being returned to the Elbow River. Most residents are smiling, although concerned opponents decry the precedent of dumping the residue directly into the Elbow — a discharge that could be more detrimental than the septic systems it replaces.

Map of the Elbow River Alluvial Aquifer, from Highway 22 to the Glenmore Reservoir (Government of Alberta)

Today, the Bragg Creek area is home to over 2,400 people, most of whom live on the scattered residential acreages surrounding the hamlet. Beginning with Jake Fullerton's cottage lots on the Elbow in the 1930s, Calgarians have been attracted to this area for its recreational opportunities. The tourist economy became firmly established following World War II and by the 1970s, low-density residential subdivisions housed many who commuted to Calgary daily, but enjoyed living in this country setting. The landscape is relatively open, interspersed with extensive areas of low-intensity agriculture and old growth forest. The hamlet's commercial core supports not only the area residents, but is heavily oriented to national and international summer tourists. And pressure for further residential development continues to grow.

Meanwhile, the environmental movers and shakers are not the only well-organized activist cell in the hamlet; there are others, covering the performing arts, senior citizens, and small business. When their historic community centre burned down in 1998, Bragg Creek residents immediately conceived a bigger vision. They raised funds and built a much grander, multipurpose structure to house theatre groups, art shows, artisan fairs, music performances, wedding receptions, and the like. When the Centre designers did not agree to meet the needs of the senior citizens, the latter group said "Humphh," raised their own money and built their own rustic, comfortable Snowbirds Seniors Centre. Now their Seniors Housing Society has approval to build a multifamily seniors housing project adjacent to the Snowbirds centre. Bragg Creek seniors – another force to be reckoned with!

Given its tiny hamlet population, the impact of Bragg Creek on its hinterland is disproportionate in the extreme. It has survived fire, a flooding river, groundwater contamination, and many industrial demands on its recreation lands, and it will continue to do so. It is this feisty pioneer attitude that sets it apart. As we head back down the road toward the big city, we mull over what issues we should take a stand on. There are many.

Part 3.

To Plains and Metropolis

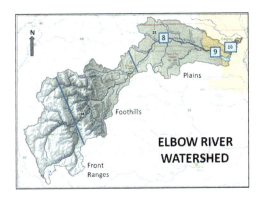

(map courtesy Bow River Basin Council)

Chapter 8.

Horses, Hunters and Homesteaders

"Wherever man has left his footprint in the long ascent from barbarism to civilization, we will find the hoofprint of the horse beside it."
—JOHN TROTWOOD MOORE

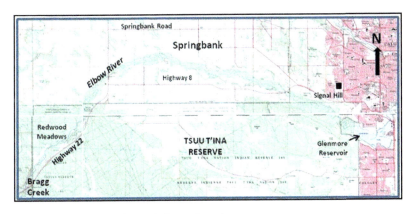

Map of the Elbow River, from Bragg Creek to the Glenmore Reservoir

I was digging in my vegetable garden when I found a rusty horseshoe buried in the dirt. *[Who could have been riding through this field?]* My interest piqued, I was off on an investigation of this watershed's quadrupedal inhabitants. In no time at all I found myself standing beside a thousand pounds of solid horseflesh, looking up at the towering chestnut beast with awe, ready to climb on. Surrounded as I am by a deep-seated and pervasive horse culture in my watershed neighbourhood, I decided I should learn to ride. So here I am at a Bragg Creek riding stable, about to fulfill my dream — all in the name of research.

The stories, the old photographs, the drawings, the books about the watershed's history all reference this background element — the horse. From the wiry tribal bison-hunting ponies, to the draft horses hauling homesteader carts across the prairie, to the skillful ranch horses working cattle on the range — these beasts had a significance in the watershed I had not suspected.

Below Bragg Creek amid the treed foothills, the Elbow watershed opens out onto a broad ochre-and-green plain centred on the meandering river: this is the central watershed. Here amongst the hay fields, pastures and residential acreages, it is hard to imagine this landscape without horses. But there was such a time. About 15,000 years ago, the equine descendants of the little dawn horse *Eohippus* disappeared from North America, possibly over the Bering land bridge. It was not until the 1500s when the Spaniards arrived in the New World that the modern horse returned to its ancestral home. By about 1750, horses had been acquired by tribes in the vicinity of the Bow and Elbow rivers, transforming hunting, warring, transportation, recreation and wealth; they were such valuable commodities that horse thieving was rampant.

The history of people in the Elbow watershed began 10,000 years ago, when the first *Homo sapiens* appeared on foot, hunting large mammals along the receding ice edge. When the Ice Age mastodons, mammoths and early horses disappeared, these nomads survived by hunting bison and other animals in the area. By 8,500 years ago, the Elbow watershed was a rugged, peaceful area, covered with deep snow and ice in its upper reaches, thick green forests in the foothills, and tall billowy grasslands in its lower reaches. The aboriginal hunters had separated into tribes or nations, each occupying a broad territory east of the Rocky Mountains over the ensuing centuries.

Eventually the Blackfoot Confederacy (consisting of Peigan/Piikani, Blood/Kainai, Blackfoot/Siksika and later Gros Ventre/A'ananin) ranged over the foothills and plains of what is now southern Alberta. Shaggy-haired, humpbacked bison (buffalo) also roamed the plains in massive herds and wintered in wooded foothills valleys, providing food, tools, clothing, fuel, bedding, and hides for tipis. Five hundred years ago, as the Spaniards landed in the Caribbean and Jacques Cartier established colonies in eastern Canada, the Sarcee/Tsuu T'ina (meaning "many people") lived to the north as part of the subarctic Dene or Beaver people, and the Stoney/Nakoda people occupied land with the Assiniboine nation far to the east.

The climate was colder than today; winters were severe and mountain glaciers had once again advanced down their valleys toward the foothills in the beginning of what was later termed the Little Ice Age. The upper valley of the Elbow watershed was snow-covered year-round, with the headwater glacier extending down the mountain into the present-day Elbow Lake basin. Below it, the icy Elbow rushed past snow and rock toward the dense forests below, which showed scars of previous fires but were otherwise pristine. Bears, bighorn sheep and mountain goats roamed freely in these upper reaches. In the foothills, Peigan hunters tracked deer, elk and moose along familiar forest trails, near encampments of several tipis scattered in foothills meadows. The human imprint on the watershed was negligible. Within two centuries, however, this elemental indigenous scene would change drastically as two cultures "discovered" each other. Life for the native peoples would never be the same. And neither would the watershed.

In 1670, the Hudson's Bay Company (HBC) was granted a charter to trade in Rupert's Land (Western Canada), and Europeans made their first appearance in the west. Thus began intensive fur trading, in which the newcomers depended entirely on local aboriginal tribes to provide not only furs, but also geographical knowledge, medicines and food. This arrangement worked relatively well for nearly a century, until the Europeans brought guns and smallpox to

Eastern Decisions – Western Impacts

In the latter half of the 19th century, major events in Eastern Canada changed the face of the West forever. In 1867, the Dominion of Canada was formed, and three years later, it purchased Rupert's Land from the HBC. Present-day southern Alberta was surveyed into precise townships and the Dominion Lands Act (1872) encouraged settlement of the West by offering homestead land practically without obligation.

However, with the HBC no longer in control of the region, confusion about the responsibility for law and order reigned. When authorities in Montana clamped down on the illegal sale of liquor in 1869, American traders flocked north across the Medicine Line into southern Alberta to acquire furs and established Fort Whoop-Up as a trading post, using whisky as their main trade item. In response, the North-West Mounted Police (NWMP), formed in 1873, based themselves at Fort MacLeod and set out to clean up the trade that was having such a devastating effect on the native population. By 1880, to make matters worse for the First Nations, the great bison herds had been eliminated through rampant slaughter. ➔

---

the west. Firearms revolutionized aboriginal hunting and warring practices. But the smallpox was devastating: over the next century, it would return three times with a vengeance and decimate native populations. The Tsuu T'ina population declined from a high of 800 in 1836 to only 100 just 30 years later.

Throughout the eighteenth century, the Elbow watershed saw sporadic First Nations and European traffic but nothing in the way of permanent settlement. In 1787, the great explorer and map-maker David Thompson passed through the Elbow watershed, sharing a Springbank-area winter camp with the Peigan who controlled the whole watershed at the time. Thompson was followed by Peter Fidler, another noted explorer, and much later, by Captain John Palliser. The Stoneys were moving west to the foothills and the Tsuu T'ina arrived in the area of the Bow and Elbow rivers to join the southern Blackfoot confederacy as nomadic bison hunters. Plains Indian historian Hugh Dempsey describes them as equestrian Indians, indicating the special value of horses in their lifestyle, and "the bravest tribe in all the plains."

Following the explorers, entrepreneurs began arriving in the watershed in great numbers. In 1871, enterprising Fred Kanouse, a former Montana sheriff, built a short-lived trading and whisky-trafficking post on the Elbow River, five kilometres up from its mouth. Fred was not the only entrepreneur in the Elbow watershed at this time. In 1873, Sam Livingston established his first trading post not far from Bragg Creek where a Stoney Indian trail forded the Elbow and where the Blackfoot often camped for the winter. About the same time, Father Constantine Scollen built the first church in southern Alberta — the Roman Catholic Our Lady of Peace Mission — near the same ford. Their commercial and religious enterprises benefited from the regular passage of the local Tsuu T'ina and Stoney tribes. Meanwhile the NWMP, arriving at the confluence of the Bow and Elbow rivers in 1875, built Fort Calgary, firmly establishing the presence of law and order in this area. An influx of Europeans was imminent, so pressure built to "settle" the First Nations people (many of whom were now sick and starving) on specified reserve lands.

Treaty Number 7 was negotiated with the Siksika, Blood, Peigan, Tsuu T'ina and Stoney peoples in 1877. The First Nations considered this a peacemaking treaty, in which they downed their weapons and shared the land; the government, on the other hand, recorded this as a land surrender. Included with the Blackfoot and Bloods on reserve land at Blackfoot Crossing on the Bow River east of Fort Calgary, the hunter/gatherer Tsuu T'ina resisted and left to carry on with their nomadic bison hunting until all bison were gone and they were starving. Miserable back on the Blackfoot Reserve and led by a determined Chief Bull Head, they relentlessly demanded a reserve of their own, even sending a petition to Ottawa. Finally, in

→ The Indian Act of 1876 then gave the federal government extensive control of the Aboriginal nations, their land and their finances. The reserves were established so that the native peoples could support themselves with agriculture on their own land, but they were in fact consigned to the status of minors and subject to the control of government agents in every aspect of daily life. The interaction of the two cultures had been reversed; beset by smallpox, alcohol and the disappearance of their life staple, the bison, the First Nations, including the Tsuu T'ina, were now dependent on the Europeans.

1883, they were given their present Reserve land (280 square kilometres straddling the Elbow and Fish Creek watersheds). There they learned farming while maintaining their hunting lifestyle on Crown lands in the mountains and foothills to the west. They had now become permanent residents of the Elbow watershed.

Tsuu T'ina emblem, in which the stretched beaver pelt represents their Beaver ancestry, the warbonnets the separation story of the Athapaskan and the Tsuu T'ina people, the peace pipe "peace with all people", and the broken arrow, "no more wars"

After two decades of economic hardship, disease and demoralization on the new Reserve, life for the Tsuu T'ina began to improve. Farming and ranching became more successful, and products like hay, berries, wood and trees were sold to the residents of nearby Calgary. After 1900, their Reserve land became, and has continued to be, a sought-after commodity. The Nation leased land to the government in 1903 for a military camp, sold about 240 hectares to the City in 1931 for the Glenmore Reservoir, and leased additional land in 1952 to the Department of National Defence. Both military leases have since been terminated and the lands returned to the Nation. The eastern third of the Reserve is now surrounded by the City on three sides, creating land use and transportation pressures. After lengthy negotiations between the Nation and the City for land to complete a municipal ring road, a deal was made in 2013. The planning involves realignment of the Elbow in the Weaselhead area, creation of two large stormwater ponds, and considerable transportation and utility corridor construction through that area. As part of the deal, the Nation takes over 2,000 hectares of forested Crown land northwest of Bragg Creek.

Spirituality has tied the Tsuu T'ina closely to their environment. With the Sun (the source of heat and light) as father, Moon as mother and Morning Star as their child, the Great Spirit is recognized in all things, including water, trees, earth, animals and birds. The Sun Dance, an important religious ceremony, was carried out for years on the north side of the Elbow in the Weaselhead area until outlawed by the government in 1876. Farther west in the Elbow watershed, Moose Mountain is another place of spiritual significance to the Nation (its subalpine fir (Sweet Pine), for example, was used in the Sun Dance and other rituals). Water as one of the four primary elements (along with fire, wind and earth) is reflected in many traditional stories. The respect of the Tsuu T'ina for the spirituality of water, and for the environment as a whole, is just one reason why they are such valued partners in the Elbow watershed.

Today, the Nation is governed by an elected Chief and 12 councillors, and numerous successful economic enterprises have been established in the watershed, including Sarcee Gravel Products. Wolf's Flat Ordnance Disposal, the only native-owned such company in North America, has done international ordnance disposal work in Kosovo, Panama and elsewhere. The Tsuu T'ina's expertise with horses was a natural advantage in taking up ranching/mixed farming, and continues to provide an independent income for individual band members.

Tsuu T'ina-owned and -operated Grey Eagle Casino and Hotel, 2014

On the banks of the Elbow not far from Bragg Creek, the Nation established the townsite of Redwood Meadows in the 1970s, including a 350-lot housing development, with its own water treatment system and an adjacent golf and country club. An annual All-Indian Rodeo, called The Greatest Indian Celebration in Canada, is held nearby. In 2007, the Grey Eagle casino complex was built on the eastern edge of the Reserve, adjacent to the City of Calgary, generating significant funding for their First Nations Charity; seven years later, a massive hotel and an entertainment centre were added. Dempsey calls the Nation "one of the most progressive and successful in Western Canada."

The Reserve sits on the western edge of the great Canadian Interior Plains, the vast physiographic region sloping ever so gently from the foothills across the prairies to eastern Manitoba and the Canadian Shield. On top of the deeply-buried sedimentary rock layers lie glacial deposits, heavily eroded by ancestral rivers flowing out of the mountains. Vestiges of these ancient rivers are now seen in the Elbow and the Bow, and within this part of the watershed, remnants of the former, higher plains are found in knolls and ridges, such as Towers Ridge, Bullhead Hill and Signal Hill.

In this undulating landscape, the aspen parkland ecoregion forms an ecological and climatological transition between the foothills boreal forest to the west and the prairie

---

The Amazing Aspen

The ubiquitous aspen poplar or trembling aspen (*Populus tremuloides*) is the most widely distributed tree in North America. Named for its long-stemmed quaking leaves, it grows rapidly to 25 metres tall, and is mostly propagated from sprouts within its own root-based colony. Each genetically identical aspen clone, or grove, has its own characteristics (such as its time of leaf-out in spring) and it is one of the largest living organisms on earth, sometimes containing hundreds of trees. Gadd describes the aspen as the world's oldest known living thing (possibly millions of years old). Aspen respond to disturbances such as fire with characteristic rapid regrowth, but each tree only lives about 50 years, typically succumbing to the shade provided by successional spruce and pine canopies.

Aspen have been used for centuries for medicinal and other purposes. Salycin, an ingredient much like that in today's aspirin and found under the black-scarred, greenish-grey bark, was extracted by First Nations people to treat various ailments. White powder from the bark itself, developed by the tree as protection against ultraviolet radiation, acts as a sunscreen and stops bleeding in wounds; the inner bark can be boiled into a cough syrup. The early ranchers and settlers valued the aspen, along with the balsam poplars and spruce usually found here in river valleys and coulees, for building homes and barns, for firewood and for fencing and corrals. Soft, non-splintering aspen wood is used today for bleached kraft pulp, some solid wood products, oriented strand board (OSB) production, and even chopsticks. ➔

grasslands to the east. Here the Elbow is flanked by grassy meadows and clumps of aspen mixed with spruce and balsam poplar. Long winters are broken up by welcome warm chinook winds, summers are warm and too short, and sunny and dry are applicable weather descriptors year-round. Some of the largest remnants of the original aspen parkland in the watershed are perhaps found on the wooded Reserve hillsides, since most parkland has been cleared for agriculture or for residential development.

→ With their strongly pumping root system, aspen are perfectly suited for phytoremediation — the plant-based cleanup of environmental pollution in water, soil or air. They help to control or remove organic contaminants (like explosives) from soil or groundwater, or to decontaminate liquid percolating through landfills. Aspen twigs, buds, bark and foliage provide food for hares, mice, moose and elk, and building materials and food for beaver. And, beyond the practical, aspen are simply appreciated here for their brilliant yellow autumn contrast with the cobalt-dark spruce.

Aspen mixed with spruce beside the Elbow River in autumn

The new Dominion government was desperate to populate the west. Although the Dominion Lands Act had been passed in 1872, a general recession and confusion over lands actually available delayed the major influx of settlers into the area. In 1881, the government offered 21-year grazing leases at one cent per acre per year to promote ranching. Eastern investors quickly established four huge ranches, controlling over 40 percent of the land and cattle in southern Alberta, but the era of the giant cattle ranches of 100,000 acres or more in the region prospered only until about 1910.

By 1883, the Canadian Pacific Railroad (CPR) had reached Fort Calgary and the process of western settlement could begin in earnest. The freed-up open grazing lands led to a second wave of Alberta sodbusters after 1900. In the Elbow watershed, homesteader lists show that available quarters within about 30 kilometres of Calgary were taken up early (before 1890), after which a steady flow of settlers took up more land until about 1910. *[John Boucher homesteaded the quarter my acreage is on in 1905. Hmm, did the old horseshoe come from his farm?]*

Before 1880, Elbow valley homesteaders were few, but after that they regularly arrived on horseback, in oxen- or horse-drawn wagons, and later on colonist trains. Men came from the 1885 Northwest Rebellion in Saskatchewan, from farms in eastern Canada, from the United States and from Europe, often with families in tow. Anxious to find their homestead quarter and get started, they set out from the western limits of the little Town of Calgary at 14th Street West for the wide-open countryside.

A narrow dirt track, leading west-southwest, near the route of present-day Richmond Road, skirted below the higher ridges to the north (Signal Hill today) and above the densely wooded Elbow River valley through what was aptly named the Spruce Vale district. Looking south, the homesteaders could see the Chipman Ranche buildings and contented grazing horses and cattle on the spring-green pastures beyond. To the west was the meandering river (identified by its tall spruce forest), scrubby aspen- and shrub-covered slopes, lushly grassed low-lying areas, and towering snow-draped mountains stretching across the whole western skyline. With little visible sign of occupation ahead, aside from T.K. Fullerton's small Spruce Vale Farm near the Elbow, it must have been a jaw-dropping sight!

Horses, Hunters and Homesteaders

Springbank homesteader's shack (Glenbow Archives NA-1241-1155)

About 10 kilometres from town, after two or three hours of jouncing, the rutted track divided. The potholed right fork (the Springbank Trail) wound bumpily through the aspen and grasses on the main terrace above the river, leading west-northwest along the path of today's Lower Springbank Road. The left fork (the Elbow Valley Trail) dipped down a gentle slope into a sombre spruce forest to meet the Elbow itself at a suitable location for fording the river and its several wandering distributaries (about a kilometre upriver from today's Twin Bridges). This mucky track continued along the floodplain for a couple of kilometres before a steady climb up to the height of land south of the river, about where Elbow River Estates is today. The way west now spread out before the traveller, as the little trail, not much more than a cart track, led over gently undulating terrain above the river. *[Maybe this is where a horse threw its shoe, since one rough map shows that the track passed by my acreage.]* Then it wove down onto broad, willow-covered flats, headed toward the foothills. Even a greenhorn could not miss the potential for ranching in this spectacular part of the watershed.

Coyote: The Clever, Crafty *Canis*

While wolves pose no threat in today's watershed, coyotes are an ever-present reality. Although rightly accused by ranchers and farmers of killing young livestock and stealing chickens, they are admired for their iron stamina, their survivability as a species, their sense of humour and their intelligent use of group behaviours to achieve their goals. This prairie or brush wolf (*Canis latrans*, or "barking dog") is the consummate adaptable survivor. Versatile coyotes hunt singly or in groups, at night or in the day, are omnivores (eat both plants and animals), eat fresh meat and carrion, and have adapted to many environments. In the Elbow watershed, they are found in grassy areas and open woods everywhere from city parks to the subalpine and even above treeline in the mountains. Coyotes are smaller and lighter-coloured than wolves, with senses of smell and hearing much better than a domestic dog. With long bushy tail carried low between their hind legs, they can run nearly 70 kilometres per hour over short distances. Conscientious mates and parents, they protectively burrow their dens into steep banks, under rock ledges, or on shrub-covered slopes. →

The first homestead quarters to be settled were grouped along these trails and near the river, with the rest filling in as more homesteaders arrived. A homesteader in this area started in a tent, a sod or log shanty, or a tarpaper shack. Basic furniture was crafted from the poplars by the river, and mattresses stuffed with straw. Provision of safe drinking water for the family and the livestock was a primary concern. The abundance of good water made the Elbow valley a favoured area for settlement. The river provided the most immediate, reliable year-round source. In the Springbank area (originally called Spring Bank, after two large springs found on hillsides to the east), many homesteaders hauled spring water in wooden barrels mounted on horse-drawn carts, and captured valuable rain water from building roofs. Wells were painstakingly dug to access groundwater and later, windmills raised wellwater for ranch and household use. Water was then, as now, a precious commodity in this dry plains area.

Ranching in the Elbow watershed was never part of the giant-ranch, open-grazing scene that dominated the foothills and open range to the south. From the outset, the people that settled in the Elbow valley were rancher-farmers, who homesteaded on their initial quarter and added small grazing leases for their manageable herds when they could afford to. Until they built their barns, fenced pastures and corrals, and planted hay and grains, their horses and cattle ranged in the open. Operating in this way, they were able to survive the threats to their livestock and their own survival and prosperity: the severe winters, drought, prairie fires, wolves and coyotes.

Beset by a host of such wildlife and climate challenges, not all the homesteaders in the watershed toughed it out and stayed; their lands were often picked up by neighbours. Farmers could thus build up their holdings and prosper, and many farms grew to between 200 and 1,600 hectares. Horses were key to their farming and ranching success. Even early on, heavy horses (particularly Clydesdales, Percherons and Belgians) were invaluable, pulling plows for sod-busting, breaking up clods with harrows, and pulling haywagons and binders. Lighter saddle horses were used for rounding up cattle, carrying children to school, and other chores requiring more mobility. By the 1900s, there were also many working horses in Calgary and in small towns, for policing, for pulling delivery wagons, and for many other jobs. Horses were bred for this growing market, which included not only a stream of new settlers, but also the Klondike gold prospectors, the Mounties and the logging industry.

→ During early settlement, coyotes did not bother horses or cattle much but took sheep and poultry, so hundreds were hunted and killed to protect the livestock. Alberta had a bounty on coyotes from about 1918 until 1948. From 1952 to 1956, an intensive strychnine campaign destroyed an astounding 170,000 coyotes. Even so, throughout the 20th century their North American numbers have increased. Since the 1950s, coyote control has changed from offensive to defensive — from poisons to electric fencing, guard animals, den removal and other more humane control techniques. Attitudes, particularly urban ones, have further altered to regard the coyote as encroached-upon, with opportunities for contact prevented where possible.

Most non-rural people are familiar with the coyote in a different sense, namely Wile E. Coyote of cartoon fame. Long before cartoons, however, Coyote was a mythological character in many First Nations cultures. He was typically a clever trickster, but also the creator (Old Man Coyote in Plains Indian myths). Coyote is also represented in well-loved children's literature, like Thornton W. Burgess' *The Adventures of Old Man Coyote*, or Thomas King's *Coyote Sings to the Moon*.

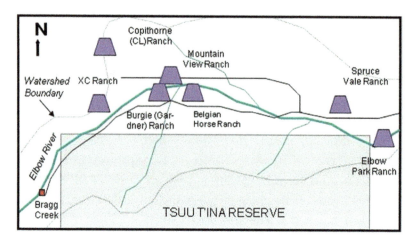

Map of early ranches in the Elbow watershed

By 1900, seven important ranches had been firmly established in the amenable lands of the Elbow watershed. These were not the vast open-grazing ranches that operated to the south, but smaller horse and cattle operations. The first horse ranch in the watershed was the Chipman Ranche, established about six kilometres from Calgary in 1883 by a group of Halifax investors, headed by the Chipman brothers. It became a celebrated supplier of Percheron heavy horses, a profitable business carried on by Richard George (RG) Robinson after his purchase of the ranch in 1888. RG took over management of the renamed Elbow Park Ranch, which covered much of the area of today's Lakeview community, Glenmore Reservoir and Mount Royal University. He built a large comfortable ranch house on a bench overlooking the Elbow River and played an important early role in developing ranching within the watershed.

RG was the first to ship cattle from this region to England, and to import heavy Shires and Clydesdales, and thoroughbreds. Even with a horse herd of 2000, he could hardly keep up with the demand for horses. A big thinker, RG expanded the horse and cattle pasture area by obtaining additional quarters further west near Pirmez Creek and at Bragg Creek, Springbank and elsewhere. Over the years, many important visitors (including two governors-general) were entertained at Elbow Park Ranch with riding and roping demonstrations, chicken and trap shooting, and other ranch-related activities.

Meanwhile, Thomas (T.K.) Fullerton's Spruce Vale mixed farm-ranch included cattle, horses, pigs and chickens. Another visionary, he soon expanded into logging in the Bragg Creek area to serve the Calgary market, eventually homesteading there with his sons.

Captain Meopham Gardner, ready for a polo match, about 1906 (Glenbow Archives NA-1942-1)

In the spring of 1886, Captain Meopham Gardner, twice wounded in the Northwest Rebellion, came west with his family to settle on the luxuriant grassy flats at Pirmez Creek beside the Elbow. There, access to water and abundant grass suited his intention to build a cattle ranch. And build Burgie Ranch he did, leasing additional pasture on the nearby Tsuu T'ina Reserve for his cattle and horses. Enthusiastic community members, Captain and Mrs. Gardner also took in paying guests (often European nobility), ran the Pirmez Creek Post Office from 1913 to 1938, and were avid

Water Woes and William Pearce

In this part of the watershed east of the foothills, annual precipitation is relatively low, at about 45 cm annually, compared to 64.5 cm in the foothills. To the early settlers, irrigation of pasture and croplands was thus a subject of some interest. In 1894, the North West Irrigation Act was passed by the federal government, formally recognizing water as a community resource, to be managed in the public interest. It reserved all water rights in the name of the Crown, establishing the foundation for contemporary water law in both Alberta and Saskatchewan. Topographic and irrigation feasibility surveys in the West were immediately initiated. The intrepid Dominion surveyor A.O. Wheeler mapped much of the Elbow watershed to assess the foothills as a source of water for irrigation further east on the plains.

In 1893, even before completion of the 1896 Wheeler map, William Pearce, a well-known Calgary engineer and surveyor who had drafted the North West Irrigation Act, promoted local irrigation. Investors formed the Calgary Irrigation Company and began construction of many kilometres of ditches in the watershed, starting near the Elbow River on the Bragg Creek trail, extending across the flats in the Milburn Creek area, southeast through the Tsuu T'ina Reserve, and on into the town of Calgary. Bad timing, however. As historically occurs in this climate, the dry years were followed by a series of wet years in which ranchers were not interested in buying water, and the Calgary Irrigation Company went bankrupt in a year. →

supporters of the early Scouting movement. One of the first Boy Scout camps in the area was set up on their land at the location of today's Camp Gardner on the Elbow River. While Mrs. Gardner wrote botanical articles for local newspapers, her husband captained the Elbow River Polo Team, went bird hunting and participated in horse shows. Taking advantage of a popular sport on many ranches, he organized a Hunt Club for hunting the abundant coyotes (foxes were too scarce). Tally-ho!

Directly across the Elbow, Louis Napoleon Blache populated his Mountain View Ranch with horses, cattle and hogs. J.A.W. Fraser's XC Ranch bred heavyweight hunter horses for the Mounties and ran Hereford, Shorthorn and Angus cattle. On the west edge of the watershed, the Irish Copithorne brothers — Richard and John — established a mixed-farm homestead, and later a sawmill. Their enterprise eventually grew into a 60 square kilometre cattle ranch in Springbank and Jumpingpound — the largest ranch in the watershed by far (20 times the area of the City of Calgary today).

Raoul Pirmez came to Calgary around 1900, with financial means and a serious interest in horses. He purchased three quarters of land on the Elbow, just east of the Gardners' Burgie Ranch; his Belgian Horse Ranch bred prize-winning Belgian draft horses. His fancy European ways, such as flying his family's "colours," were initially the object of amusement for the local cowpunchers. Like the other successful operations in the watershed, the ranch also supported other activities; the first Pirmez Post Office was opened in his home in 1910, and a federal fish hatchery was established on his land near Pirmez Creek. The creek bearing his name carries his legacy in the area, although he only operated the ranch for a decade before moving into Calgary as the Belgian Consul.

Meanwhile, considerable homesteading activity was taking place north of the Elbow in Springbank. After 1885, the Young families swarmed into the area from England via Ontario. Brothers James, William and Thomas Young all homesteaded around Springbank Creek. On a cold November day, their families arrived on a colonist train with livestock, household furniture and supplies, tools and machinery, a grand piano and a new wagon. They reached Springbank just in time to experience the most severe winter in decades. Fortunately, log houses and a barn, albeit doorless, had been built before their arrival, but the larger horses had to spend the long, bitter winter outside and the chickens barely survived in a chilly covered hole in the ground. But persevering through blizzards, intense summer heat, ruthless prairie grass fires, drought and hail storms, the Young families promoted community development, donated land for a school and a church, imported the first steam threshing machine in the district, operated the post office and a store for two decades, started a short-lived

→ Pearce also had the wisdom to promote the preservation of the prairie river headwaters in the Rocky Mountains as pristine areas free of intervention by logging, mining, settling and grazing interests. This resulted in the creation of the Foot Hills Forest Reserve in 1898. A man of foresight.

cheese factory to supply the Calgary market, and provided midwifery services. Their descendants remain in Springbank, and their community involvement has provided a standard still emulated today.

The early 1890s were exceptionally dry years and irrigation of pasture and cropland looked like a godsend to local ranchers and farmers. Visualizing the benefits of irrigation to his Burgie Ranch, Captain Gardner, with federal permission, constructed an irrigation system on his and adjacent land, using water from a tributary of the Elbow (Pirmez Creek). This system remained in use for many decades. North of the Elbow, beginning in 1896, the Springbank Irrigation Company built about 60 kilometres of canal, initially irrigating 8,500 hectares of farmland.

As more settlers arrived in the watershed west of town, community amenities began to appear, in part to combat the loneliness of living on widely separated farmsteads. In the Spruce Vale district in the 1880s, a tiny log school built by homesteaders for the children of the few surrounding farms, also served as a church; in Springbank, a similar practical solution. By 1909, the Springbank Methodists and the Union Church each had their own buildings. Amenities like rural telephone service and twice-a-week mail service were brought into the watershed.

Three dozen students outside Springbank School, 1905, all but a few of them with a horse (Glenbow Archives ND-42-1)

Most children had a long trip to school on horseback, often two or three siblings on one horse, in all kinds of weather. In 1931, the Dawson Hill School was built south of the river to shorten the trip for many of the children. *[Near my acreage! Did the horseshoe come from one of the children's horses?]* The one-room schoolhouses also hosted community meetings, dances, concerts and other events in the early days. Community halls were constructed to become a third focal building in which people gathered: Springbank Hall (1910), Cooper Memorial Hall (1928) and Elbow Valley Hall (1929): all are still in use today. While the settlers worked hard to establish themselves on their land, they also made time for recreation. Fishing in the meandering Elbow or a Saskatoon berry-picking picnic involved the whole family and had the added benefit of providing food for the table. In addition, clubs were organized — a Rifle Club, a Polo Club, a Hunting Club, a Pony Club, Ladies Aid and Missionary Societies. Rodeos, horse races, gymkhanas and dances were all popular events.

The tradition of the family ranch in the Elbow watershed and surrounding region has continued uninterrupted since homesteading days. Several of the original families, through hard work, forward-thinking business practices, and accumulation of land and stock, have remained as dominant forces in the watershed throughout the twentieth century and into the twenty-first. For example, by 1968, Rocky View County land ownership maps show that the descendants of the pioneer Young families still owned three sections of valuable Springbank ranch/farm land. South of the Elbow, Meopham Gardner's descendants operated nearly 10 square kilometres of prime ranchland as the Gardner Cattle Company. The Copithorne holdings had spread out over nearly 47 square kilometres, while the Fullertons held more than three sections. Then the original family names began to disappear from the maps and the size of the old ranches decreased markedly, as portions of their properties were sold to developers.

In Springbank in the 1970s, small acreages owned by city folks began to pop up like gophers. South of the river, large-scale conversion of farm and ranch land started later but ramped up quickly after the year 2000. Much of Highland Stock Farms, an Elbow valley cattle breeding ranch established in the 1930s, is now a high-priced suburb, while the Gardner sections to the west are currently being promoted for a large housing development. The trend is obvious. Every ranch in the watershed has declined in area from 1968 to 2011, while the developed land or land purchased for future development has skyrocketed from one percent in 1968 to over 30 percent by 2011. If a conservative estimate of developer-optioned land is included, their share rises to 50 percent. Such is the fate of scenic agricultural land sitting too close to a booming, voracious metropolis like Calgary. Sometime in the near future, most of this plains area of the Elbow watershed may be covered not just with acreages, but with paved roads and housing developments, despite pleas to "keep it natural." Natural-world enthusiasts hope that the municipal planners ensure that any development occurs in a controlled sustainable way.

The conversion of the land and its loss to agriculture is not the only worrisome issue, however. Water has been, is and will continue to be, critical to the viability of life in this central part of the watershed. Today, this is the most populated part of the watershed outside the city limits, almost entirely within the jurisdiction of Rocky View County. Here water is a bone of contention, as more rural residential and commercial developments are planned and developers are now required by the County to provide solutions for provision of water and disposal of sewage before their proposals are considered. Increases in phosphorus and nitrogen levels in the Elbow have been discovered, linked to fertilizers, detergents and human and animal wastes. As densities increase, the effects on water quality here and downstream will also increase: more pesticides, herbicides, manure, road salt and septic systems all impact the alluvial aquifer and the river. This seems to be headed in a dangerous direction, unless very carefully managed.

Horse pasture at a riding stable in the Elbow valley, surrounded by houses

My small acreage has its own septic system and potable water is provided through a community water co-operative from wells accessing the Elbow River aquifer. This system, similar to that used by other communities in the watershed, works relatively well. Larger acreages and farms have their own wells and septic systems. At present, there is no other option for piped potable water or septic services. The City and the County are at war on this issue, the City having excess capacity in its water license but declining to make water available to additional developments in the adjacent County. Private firms are using co-opted water licenses to sell expensive potable water to new communities in some locations. One large new residential development is currently buying its water from a private provider and trucking its sewage to a Calgary treatment plant, a far-from-ideal situation. As population densities in the watershed increase, the importance of this issue grows accordingly, and no solution is immediately apparent.

The County and Alberta ESRD are attempting to get a handle on it, though. They are limiting new development on the alluvial

Frustrating FITFIR

Demand for water from a finite and even declining supply has grown rapidly with population and economic growth, and the current water allocation system poses a problem. Since 1894 in Alberta, Canada's driest province, all potential water users had been required to apply to the government for a licence to divert water from any surface watercourse (and in a small number of cases, from groundwater), and to make a one-time payment based on the volume of water involved. The 1931 Water Resources Act officially made all water the property of the provincial government. Based on estimates of its sustainable yield, the water licences allocated the available water supply. Allocation was based on a First in Time, First in Right (FITFIR) system, in which the first licences granted had the first priority, particularly in times of water scarcity. The water licence was tied to the land, passing from owner to owner.

In 2006, realizing that increasing demand on a decreasing supply was a problem, the provincial government imposed a moratorium on new water licences in the Bow, Oldman and South Saskatchewan river basins, although licences could be transferred (traded) between users. This protected the current water licences, and restricted the amount of water available. Accordingly, the 132 licenses on the Elbow River were temporarily frozen. Most of the Elbow's fully allocated water supply was being used for municipal purposes, with a small allocation for agriculture and for environmental sustainability; however, frustrated land developers continue to be hungry for additional supply.

aquifer, amending bylaws and procedures to protect riparian areas, wetlands and the aquifer, and putting other best practices in place. To their credit, the previously unopposed rampant development has been temporarily slowed by the County, until appropriate measures and plans can be put in place. This will be a continuing issue, and not only in our Elbow watershed.

Despite urban development, its associated influences and water woes, the culture of the horse carries on. Today, there is an estimated 150,000 horses within a 75-km radius of the City of Calgary (nearly half of the 313,500 horses in Alberta), putting something like 2,500 horses within the central Elbow River watershed on ranches and acreages. As one horse enthusiast puts it, "Today, when only those who like horses own them, it is a far better time for horses."

Since their early hard-working days, horses' lives have indeed changed remarkably; today, recreational riding is the most popular use for horses, followed by breeding, trail riding, companionship, ranch or farm work and riding lessons. Most of the horses in the watershed are light horses (Appaloosas, Paints, Quarter Horses, Arabians), while the few heavy horses are kept almost entirely for show. The local riding community is a friendly one, accommodating English and Western riders equally. Facilities for horse enthusiasts abound in the watershed, including Pony Clubs, a gymkhana club, 4-H clubs and an equestrian society, as well as equestrian campgrounds and trails in Kananaskis Country, riding stables, equestrian veterinarians, farrier services, community riding arenas and backcountry trail riding services.

"Feeling down? Saddle up!": an aphorism that speaks to many

Alas, my research did not solve the mystery of its provenance but that little horseshoe is nailed above my side door for luck. It won't be luck, though, but rather people who keep this watershed happy and healthy, or not. How we live our lives in this watershed, how we use the land and how we treat the water — it all depends on us to make good sustainable choices.

Chapter 9.

Warriors in the Watershed

Map of military locations in the Elbow River watershed

As the Elbow winds its way across the City limits, it is winding its way through an area that echoes with the boots of a multitude of temporary watershed residents. Since the establishment of Fort Calgary by the NWMP in 1875 to the present day, the small Elbow River watershed has had a disproportionately dominant military presence within the Calgary region, and indeed within southern Alberta.

Several of the first settlers in the western Elbow watershed, including T.K. Fullerton, Meopham Gardner and Louis Blache, had early military experience, involved as they were in quelling the 1885

North West Rebellion of the Metis. When the new century dawned, and volunteers were required for the Boer War in South Africa, the Calgary-area Lord Strathcona's Horse (Royal Canadians) (LSHRC) took up the challenge and proved themselves excellent shock troops and scouts. Equipped with 600 sturdy western cowponies and western saddles, Stetson hats (later the NWMP's official headgear), Lee-Enfield rifles, lassoes and revolvers, they were mainly range-toughened bachelors who could ride and shoot with the best of them — just what was needed to deal with the Boers.

Several settlers brought their South African conflict experience back into the watershed. Major Patrick (Paddy) Drummond of the Irish Fusiliers served in the first Boer War, then emigrated to Springbank to homestead and run the first store in the area: Drummond's "Emporium" as it was grandly called, a landmark on the Springbank Trail (where Springbank Road meets Highway 22 today). Richard Cox, who had joined the British Army at 17 as a young bugler and served in the Boer War, also settled in Springbank. He was subsequently recalled to England to fight in World War I. Leaving his wife and five children to run the farm, he returned safely to live long into his nineties. Joseph Robinson, RG's eldest son who handled the cattle side of the Elbow Park Ranch business, joined the 2nd Canadian Mounted Rifles (CMR) in Calgary at the age of 24 for a new challenge. He saw plenty of action in South Africa throughout 1901 (where his closest call was having the maple leaf shot off his hat), before returning to the Elbow valley ranch.

The brief conflict in South Africa had made the deficiencies of the 35,000-member Canadian militia volunteers glaringly apparent, and steps were immediately taken to improve their training. From 1901 to 1911, summer militia training camps were held on the Colonel James Walker estate beside the Bow River in Calgary's Inglewood district. By 1910, the first Calgary militia regiment had been formed; this 103rd Regiment (Calgary Rifles), and the origin of today's King's Own Calgary Regiment, included many Boer War veterans. Growing rapidly, the 2,000-man militia moved from Inglewood to the first of several training locations within the Elbow watershed — Reservoir Park, near the later site of Canadian Forces Base Calgary.

Sarcee Training Camp, 1915 (courtesy Library and Archives Canada PA-147485)

Very soon, though, the threat of World War I loomed in Europe, and the need for a much larger training area outside the city was obvious. In 1914, a large area of fairly level but scrubby Elbow valley land in the northeast corner of the Tsuu T'ina Reserve was leased by the military, and rapidly turned into the second largest military training area in all of Canada. The river valley provided flat areas for the camp and rifle range, and a variety of terrain, water and vegetation suitable for military exercises. Designed by military engineers, the Sarcee Training Camp included canvas bell tents, wooden administration buildings, an electrical system, and a system for bringing water from the Elbow River. By 1916, the massive camp accommodated 15,000 soldiers for basic training prior to being shipped overseas.

Sarcee Camp, by now the headquarters for Canada's Military District 13, stood beyond the 37th Street city limits, and the reach of all-weather roads, so the City of Calgary generously extended its Municipal Railway streetcar line a few kilometres to the northeast corner of the camp. This line was promoted as a means for families to visit the soldiers stationed there, for the soldiers to visit the City proper, and for supplies to be brought to the Camp on boxcars. This streetcar service continued throughout the war until 1919, when the CPR right-of-way lease expired and the tracks were removed. Other Camp services were provided in Sarcee City, a phalanx of wooden buildings north of the camp, housing a bakery, a café, a pool hall, and a theatre. In addition, Dempsey reports, prostitutes "peddled their wares from the nearby hills." As the Camp was not winterized, the "Tented City" operated only six months a year, May through October. But by the end of the war, an astounding 40,000 men had been processed through the Camp.

The nearby Elbow River was significant for the operation of the Sarcee Training Camp. First, it supplied water for the camp, a critical requirement that had to be carefully engineered as the camp was far from a city water supply. It also provided a physical challenge for men and horses to slog through in training. (Perhaps a little fishing, too?) And finally, it supplied rocks.

Stones depict the insignia of High River area 137th Battalion, C.E.F., and line walkway between tents, 1916 (Glenbow Archives NA-1617-3)

Rock art (geoglyphs) has a long tradition in the military. Geoglyphs are large-motif designs on the landscape, created either by placing stones on the surface, or by carving the design into the soil or rock. Military geoglyphs often represent the badges or insignia of specific regiments stationed in a locale; others are more general in nature. Sarcee Camp soldiers decided to create their own military geoglyphs. From the Elbow and its floodplain, stones were collected by the soldiers, carefully cleaned, and first used to demarcate roadways, sections of the camp and battalion areas. Detailed battalion unit badges were created in front of their designated areas of the camp, demonstrating pride in the unit and *esprit de corps*.

Signal Hill rock battalion numbers above busy shopping centre, 2014

The geoglyphs serving as a lasting memorial, however, are the battalion numbers found on the hillside above Sarcee Camp. In an extreme method of physical conditioning in the hot summer of 1916, the soldiers of several battalion units, marching in fours, hauled 16,000 rocks in rough sacks up from the Elbow River to a south-facing hill a kilometre northwest of the camp. The men of the 137th (High River), the 113th (Lethbridge), the 151st (Red Deer) and the 51st (Edmonton) formed their monumental battalion numbers (some almost 22m high) with the rocks on the hillside where they could be seen clearly from the camp, and whitewashed the rocks to further enhance their visibility. Then, off they went to war.

More than a hundred men from the Elbow watershed west of the city had enlisted during a 1914 western Canada recruiting blitz, with their largest number joining the 12th CMR in Calgary. The Mays, Copithornes, Youngs, Robinsons and many other Elbow valley farm and ranch families sent fathers and sons first to the Sarcee Camp and thence to Europe. These men were typically hard-working farm boys, already skilled at the basic military requirements of shooting and riding. At the turn of the century, most settlers could certainly handle a rifle or shotgun, but in the Elbow watershed this skill was taken to another level by the formation of the Elbow River Rifle Club. Organized by Captain Ernest May, who lived on the flats near Twin Bridges, the club had two ranges in the river valley — one in Springbank and one near Captain May's residence. The Club produced such superior marksmen that they won many trophies, competed strongly in national meets in Ottawa, and later used their talents in the battlefields of Europe in World War I. The results of one shoot showed that seven of the top ten marksmen were Youngs from Springbank.

When at last the Great War was over and the soldiers returned home (minus 60,000 of their Canadian colleagues), Sarcee Training Camp operations carried on, as there was now more emphasis on the local militia and their summer training sessions. From 1929 to 1945, Sarcee Camp served as the main summer training camp for Alberta. Meanwhile, additional military facilities were being created within the watershed on the edges of the city.

By the 1930s, drought and depression had hit the area hard, and a make-work project to create a new barracks for the LSHRC regiment was initiated by the Canadian Army. The result was Currie Barracks, named for

---

Signal Hill Today

For 70 years after being arranged by the Sarcee Camp soldiers in 1916, rock battalion numbers remained on Cairn Hill, the whitewash gradually fading, and the rocks overgrown by grasses and shrubs. In the late 1980s, plans for a new subdivision and a shopping centre below the hill threatened the memorial and prompted an outcry from a local architect and World War II veteran. When the land owner, the City, the provincial government and others got involved, things began to happen. The rocks were gridded, numbered and slightly relocated to a graded slope. Local military cadets undertook to refurbish the battalion rocks. The upper hillside, now called Signal Hill, was designated an historic site. And finally, a hillside memorial park — Battalion Park — was dedicated in 1991, a visible and fitting reminder of the Alberta soldiers who trained below this hill for a distant war in Europe.

the World War I Canadian commander Sir Arthur ("Old Guts and Garters" as Pierre Berton reports). It was located on 24th Street south (now Crowchild Trail), north of the then newly flooded Glenmore Reservoir, and a short ride away from Sarcee Camp. The new quarters, with better barns for the horses and with trees and grass, were luxurious compared to their previous cramped accommodations at the Mewata Armouries downtown.

Lord Strathcona's Horse at Currie Barracks opening, 1933 (Glenbow Archives NB-H-16-445)

Meanwhile, a new interest for the military was growing. A decade earlier in 1927, the Calgary Aero Club had been founded by Fred McCall, the celebrated World War I flying ace. Its first airfield was constructed on the Banff Coach Road near the northern boundary of the Elbow watershed. Southern Albertans took to flying immediately in this area of legendary clear skies, broad open spaces and nearby mountains. Only two years after its formation, the club was the largest in Canada and had moved to a new airfield in Renfrew, on the flats above Nose Creek. Over the next decade, many pilots were trained in its Gipsy Moths and Tiger Moths, several of which had been supplied by the Department of National Defence. It was in its interest, the government realized, to have a cadre of trained civilian pilots available.

Off to the Great War: The Springbank Youngs

Lt. Rex Young and wife Jean, at their wedding, 1915 (Glenbow Archives NA-3204-5)

William, James and Thomas Young had produced what the Foothills Historical Society described as a handsome brood of "fishing, hunting, shooting Youngs." Rex Young and his brothers and cousins practiced at the rifle ranges set up in the Elbow valley, and joined the Elbow River Rifle Club when it was formed. By 1910, at age 26, Rex had his own farm along the Elbow near the Sarcee Reserve. But seven months after his 1915 marriage he was in training at Sarcee Camp, just five kilometres east of his farm; a month later, he was sent overseas. After transferring to the LSHRC with some of his Elbow valley friends in order to stay in the action, Rex fought bravely in France. He was killed in December 1917, leaving his young wife alone with their two-year-old son. George, Walter and Percy, all bachelor sons of William Young, also enlisted with enthusiasm and all were soon overseas at the Front. After only a few months, the parents were notified that George was missing and presumed dead after the battles at Ypres. ➔

In 1935, the decade-old RCAF flew into town to manage the construction of Currie Field (the Calgary Military Airport) on the south side of the new Currie Barracks. By 1938, two RCAF squadrons were based there, flying Westland Wapiti and Siskin biplanes and, later, Hawker Hurricane monoplane fighters. When Canada joined the Second World War in Europe, the Calgary area was well prepared with both training facilities and experienced pilots.

Late summer and early fall in the Elbow River watershed in 1939 were weather-perfect — warm and dry under bright blue skies. The last of the hay and grains was being harvested and stored for the winter, and the cattle rounded up for their annual fall trip home from their summer pastures in the upper watershed. As the local ranchers were cleaning their rifles for hunting season, the news came that Canada had declared war on Germany and entered World War II. Local militia units immediately mobilized and nearly one hundred of the residents of the central Elbow watershed enlisted at once. But this time, the enlistees included five young women, and assignments led not only to the cavalry and infantry regiments, but also to the Air Force and even the Navy.

Cessna Crane trainers used in the SFTS in Alberta (courtesy Canadian Warplane Heritage Museum)

Warriors in the Watershed

Military operations within the watershed ramped up quickly at the start of the war. By 1940, 12,000 troops were training in Calgary and the LSHRC had moved to Winnipeg. Currie Barracks became an A-16 Infantry Training Centre for units raised in Alberta, before they were shipped overseas. The same year, the No. 3 Service Flying Training School (SFTS) opened on Currie Field (later renamed RCAF Station Lincoln Park) to train wartime pilots for the BCATP in Avro Ansons and Cessna Cranes (twin-engined trainers). Calgary was designated the headquarters of the Western Command. The hundred enlistees from the central Elbow watershed were dispersed throughout Canada, England, Europe and the Pacific to serve as drivers, mechanics, administrative personnel, medics and trainers, as well as soldiers, sailors and airmen. Back at home, their families vigorously supported the war effort and endured rationing of not only gasoline and tires, but also meat, sugar, coffee, tea, butter and alcohol. By the end of the war, at least five of the central Elbow soldiers had lost their lives and many more had been wounded. The veterans at last returned to tumultuous welcomes.

The post-war period at Currie Barracks saw LSHRC and 1st Battalion, Princess Patricia's Canadian Light Infantry (PPCLI) move in, filling the Base to bursting. By 1948, additional lands had been acquired south and east of the Barracks to provide housing for regimental personnel and their families,

→ Meanwhile, Percy wrote home about playing football, having chance meetings with old friends, and watching ducks pass overhead, musing about bagging one for supper. Along with 3,597 other soldiers, Private Young was killed amid the muck, snow, sleet and cold of No-Man's-Land at Vimy Ridge in April 1917. (One of Percy's fellow soldiers, Private John Pattison from Calgary, was posthumously awarded the Victoria Cross for his bravery in rushing a German machine gun that same day. The Elbow River's Pattison Bridge in Calgary is named for this undaunted soldier.)

The remaining Young brother, Walter, was in and out of the fighting, as a hernia and then a battle wound sent him from the trenches in Europe back to England for convalescence. During these respite times, he taught musketry and was anxious to get back into action. As he said, "I only am left to carry on for the Youngs at the Front..." He returned home to the farm in 1918 and lived quietly with his brother Frank and family, remaining in Springbank for the rest of his life.

I ponder the sacrifices of these and other families whenever I see Thomas Young's historic cabin on Springbank Road and the farms of the remaining Youngs.

Off to World War II: The Elbow Valley Matthews

Second Lieutenant Donald C. Matthews (3rd row, 4th from L) in Calgary Highlanders, 1939 (Glenbow Archives NA-2363-3)

Out on the Bragg Creek road past Twin Bridges, Charlie Matthews had purchased seven quarters of prime grazing and hay land in 1933, using the profits from his family piano business in Calgary. By 1937, he ran a thriving stock farm and was able to send his eldest son, Don, off to study Agriculture at the University of Alberta. When war was declared, 21-year-old Don left school to enlist as a 2nd Lieutenant in the Calgary Highlanders (one of the first Calgary units to be mobilized for the war). After a short training period at Currie Barracks, he was off to England and the Front. By 1944, Don had joined the 14th Calgary Regiment (Tanks) to stay in the action. His younger brother Dick, also a 2nd Lieutenant, joined the Navy and served in the Pacific. ➔

and later schools and churches. With the onset of the Korean War in 1950, Currie Barracks further expanded as Headquarters Calgary Garrison. Again, Calgary's regiments distinguished themselves in battle; the first Canadian unit to enter active warfare with the United Nations forces in Korea was from Calgary's PPCLI.

Once again, more land was needed, and an additional 380 hectares were acquired from the Tsuu T'ina Nation in 1952 to create the Sarcee Barracks and training area, adjacent to the old Sarcee Training Camp. LSHRC moved there from Currie Barracks in 1958; there were now men's barracks, a training hall, a workshop, a school and married quarters (later redesignated as Harvey Barracks, after Brigadier Frederick Harvey, a distinguished LSHRC officer). Meanwhile, the RCAF Lincoln Field was being actively used to train NATO pilots, giving this military facility international recognition.

These two decades (1945–64) were likely the heyday of military development and activity in Calgary, and certainly within the watershed. But as the City ballooned out to the west and as Canada moved to a peacekeeping role in the world, directions changed markedly. Surrounded by urban development, RCAF Station Lincoln Park was closed to aviation in 1964 and part of the land put up for sale. Mount Royal College (now University), founded in 1911, had outgrown its downtown location.

Somewhat surprisingly, it won a competition for the airfield land against the Calgary Stampede (under Don Matthews' presidency) also seeking expansion land. Mount Royal College built their Lincoln Park campus on the runways and taxiways of Lincoln Field.

Currie Barracks was renamed Canadian Forces Base (CFB) Calgary in 1966 (in anticipation of the unification of Canadian Forces in 1968). By 1981, the 48 square kilometres of leased land in the area of Harvey Barracks and the Sarcee Training Area were no longer needed by CFB Calgary and were prepared for return to the Tsuu T'ina Nation. A complex 20-year clearance program began to clean the soil, vegetation and groundwater of the effects of 67 years of military training and operations. Remediation of the land — through which runs the Elbow River — involved location and removal of unexploded ordnance, cleanup of explosive ordnance contamination, and separation of waste materials. Old military housing units and other structures not needed by the Tsuu T'ina were demolished.

Two areas which received special reclamation attention were waste dumps located on the escarpment of the Elbow River. These contained metals, solvents, hydrocarbons and other waste materials, all with the potential of seriously contaminating the surface water in the river and the groundwater, including the aquifer, just a kilometre or so from the Glenmore Reservoir, Calgary's major source of drinking water. Today, the area is

→ When the war was over, Lt. Don Matthews, brought his new English wife and young daughter to the home ranch by the Elbow. In the decades following, Don showed his leadership by making his father's Highland Stock Farms in the Elbow valley an international Angus and Limousin breeding success. He organized the first 4-H club in Springbank (and was later elected the Canadian 4-H President), developed the Canadian Beef Breeds Council, and served as President of the Calgary Stampede. Dick completed degrees in law and commerce, and became a successful lawyer, generous philanthropist to the arts in Calgary, and a Member of the Order of Canada. Highland Stock Farms has today moved west in the Elbow Valley to Bragg Creek and north to Olds. Much of their central watershed land has been sold to developers and is being rapidly covered with houses and pavement. Both Don and Dick might have viewed that as progress.

considered decontaminated to the level of suitability for residential use, and is home to a large and profitable Tsuu T'ina casino. A decade from now, the main area of the Sarcee Training Camp will be covered with an intricate maze of cloverleaf off-ramps, bridges and four-lane highways as part of the Calgary Ring Road, with a re-routed Elbow River beneath them. A far cry from the scrubby plain that hosted 40,000 soldiers.

Current and former Calgary-based regiments have proudly contributed peacekeeping personnel in Cambodia, the Middle East and the Persian Gulf, as well as the former Yugoslavia and Sudan. More recently, the Calgary Highlanders, the PPCLI, the King's Own Calgary Regiment (KOCR) and others have sent soldiers to Afghanistan, and to conflicts in Iraq, Libya and Mali. But in 1998 CFB Calgary was closed as part of a federal program to consolidate military facilities across Canada, and most of its personnel were relocated to CFB Edmonton. It was the end of an era in Calgary.

Military memorial and housing in Garrison Woods

Today, the Base lands are being redeveloped by the Canada Lands Company as award-winning community housing, with assiduous reference throughout to its distinguished military history and careful preservation of a few original buildings. Meanwhile, south of the old Lincoln Park main runway, former airfield hangars are being replaced by a large modular structure manufacturing operation. The consolidated Military Museums, also located on former Base lands, present a well-designed and thought-provoking display of local military history. A small active military presence in the Elbow watershed area is maintained by three naval and army reserve units and two military bands (HMCS Tecumseh Band and the Regimental Band of the KOCR).

There have been many warriors but no wars in the Elbow River watershed. Over the course of the 20th century, the watershed played host to a massive buildup of military forces, equipment and facilities for participation in four wars as well as peacekeeping initiatives. And at the end of those hundred years, the military presence is essentially limited to memorials, museums and monuments providing visitors and residents like me opportunities for remembrance. The tradition of warriors in the watershed, however, continues in the ongoing battle to preserve and protect the Elbow.

Chapter 10.

Missions, Cowboys and an Urban Thirst

Map of the urban Elbow: from the Weaselhead to the Bow

Sailing lessons on the Glenmore Reservoir — how many Calgary sailors have got their start that way? Today the reservoir hosts many little boats, white sails straining as they dash out to the buoy and back. Near the north shore the long dark dragon boats are out for some training, paddles dipping and pulling in perfect rhythm. Further west, two red canoes move smoothly through the calmer water in the lee of the shore, headed to the Weaselhead wetlands at the head of the reservoir. Blast of a horn and the S.S. Moyie turns a corner on its way back to Heritage Park. A nice summer day on the water. But recreation is not the reason this reservoir exists.

The Glenmore Reservoir was created in the 1930s to meet an insatiable need for water in a rapidly growing city. Today, Calgary, like all such municipalities, is responsible for providing safe water for drinking and other uses, as well as proper disposal of wastewater and stormwater, all to very high standards. Those regulations have been a long time coming. In early days, the water supply was not treated at all, but was taken directly from, and disposed into, the Bow and Elbow rivers. When municipal officials realized that water quality had become a major health issue, they treated the water supply prior to distribution. Much later, they also began to treat wastewater before it was returned to the watercourse. But amazingly, it was not until the 1970s that provincial regulations were imposed on both. And as water quantity became an issue, daily and annual caps were also placed on extraction of water from watercourses.

Elbow River flume repair after a flood, 1910 (Glenbow Archives NA-3496-36)

Since its earliest development, Calgary has depended on its rivers for water. By 1900, however, the Bow had become significantly polluted due to upstream activities like mining and forestry; the Elbow, with its more pristine watershed, offered a purer source of water. To better access this water, a 16-kilometre gravity feed water line, promoted by engineer and City Council member John "Gravity" Watson, was surveyed from the Elbow River west of town to a city reservoir at 24th Street and 26th Avenue SW (near today's Richmond Green Golf Course). Within two years after its 1907 approval, the immense wooden flume was completed and Elbow River water flowed into the Calgary reservoir.

A dragon boat, little sailboats and the S.S. Moyie enjoy the Glenmore Reservoir beside Heritage Park
(Photo courtesy of Robert Lee)

Only partly buried, and supplied with water from a constantly changing river course, the flume quickly became a nightmare to manage. It sagged in places and the stream bed required dredging to maintain water inflow. It frequently was choked with ice, provided turbid water during high flow, and trapped fish in the inflow screens.

But the City kept it working for nearly three decades, until a better waterworks solution could be implemented. Still wanting to access the clean and clear waters of the Elbow, the City negotiated the acquisition of land in the Elbow River valley, including the original Glenmore farm of Sam Livingston and a portion of the Tsuu T'ina Reserve. By 1933, the Glenmore Dam was completed, the new reservoir flooded, and a water plant for filtration and chemical treatment was operational, providing treated water and a measure of downstream flood control for the City. Grant MacEwan described the relatively small reservoir as "a lake so exquisitely beautiful that it won the highest praise from many international observers."

Even before the reservoir was filled, the new dam proved its value by preventing the major spring flood of 1932 from causing significant damage downstream. Over the years, its treatment plant has been expanded and upgraded to keep pace with the City's growing thirst. Today, Calgary's water supply comes from both the Elbow and the Bow and is treated at two plants: Glenmore on the Elbow, and Bearspaw (built 1972) on the Bow.

Upstream from the new reservoir was Weaselhead Flats, included in the 1931 land purchase to help protect the source of the drinking water. This 400-hectare area of critical wildlife habitat located between the reservoir and the Tsuu T'ina Reserve is today part of a three-park system encircling the western end of the reservoir. The Weaselhead Natural Area, named for a Tsuu T'ina chief, supports five distinct habitats and over a thousand species of vegetation and wildlife. White spruce forests, deciduous riverine forests, clones of trembling aspen, shrubs and grasslands are interspersed with streams, ponds and wetlands. Beaver, hare, moose, deer, black bear, coyote and lynx, plus numerous aquatic invertebrates and birds populate the distinctive habitat there. Bird watchers abound in the Flats as well! On one trip around the reservoir, where the pathway dips down into the Weaselhead, I spotted a dark brown cow moose and her chestnut-coloured twin calves knee-deep in a glassy beaver pond, munching on sodium-rich aquatic plants. Although the pond engineers — beavers — were nowhere in sight, evidence of their

Missions, Cowboys and an Urban Thirst

activity was everywhere: notched tree stumps, a large log and mud dam, and of course, the pond itself.

On the banks of the Elbow, the City of Calgary is continually dealing with beavers. These so-called pests fell prized urban trees and attempt to dam tributaries and distributaries causing flooding of public and private property. Calgary's control program deals with the beaver *in situ* rather than by relocation, placing stucco wire around trees and using "beaver bafflers" (large-diameter underwater pipes) to drain ponds through the dams.

The beaver is considered a keystone species, that is, one that has a disproportionate effect on its environment compared to its size. Beaver dams create quiet ponds with marshy edges, and tree-felling provides forest gaps and allows for shrubby growth. These ponds and gaps constitute important habitat for birds, fish, and small and large mammals. In fact, no other animal has influenced the history of Canada as has the beaver, indirectly through development of the "wilderness" by the fur traders, and directly through its modification of the landscape. Recognizing this significance are thousands of beaver-named features across the country, including Beaver Flats and Beaverlodge Trail in the Elbow watershed. And today, beavers are regaining favour in watersheds like the Elbow, where their dams and ponds act as buffers for heavy rainfall and help to capture floodwaters.

Busy Beaver Basics

These large rodents, with their much-prized thick fur pelt and flat hairless tail, are recognized worldwide as the national symbol of Canada. Before the fur trade got into full swing in the 18th century, there were one to two hundred million beavers in North America. Today, although they number only ten to fifteen million, they have been more fortunate than the buffalo in surviving the onslaught of so-called civilization. These animals are well adapted to both their land and water environments. They survive in cold water protected by their dense undercoat, self-stopping nostrils and ears, and webbed toes. Their lips close behind their incisor teeth to permit underwater chewing, and transparent membranes can cover their eyes. On land, they eat twigs, bark, seeds and leaves, and their continuously-growing incisors can fell a 25-cm aspen tree in a few minutes.

Beavers construct dams of sticks, logs, stones and mud on a slow-moving, marshy stream, creating a small pond for swimming. They then build a lodge in which to raise a family and to stay safe and warm during winter. In 2007, the largest beaver dam on record was found in Wood Buffalo National Park in northern Alberta — an amazing 850m in length. Don't underestimate your local beavers!

Upstream from the Weaselhead is a unique area in the Elbow valley: the Griffith Woods Natural Area. This donated land along the Elbow between Highway 8 and the Tsuu T'ina Reserve was opened by the City as a Special Protection Natural Park (like Weaselhead Flats to the east) in 2000. Its stand of giant white spruce is one of only two within the City, and its natural habitats sustain many species of birds, aquatic invertebrates, mammals and reptiles. A large part of its environmental significance is its function as a water and air purification system, which is important for areas further downstream. Griffith Woods is regularly used as an environmental education resource for Calgary schoolchildren.

Part of the Elbow Valley Constructed Wetland

A second educational resource in this reach of the Elbow concerns stormwater (surface runoff from precipitation and snowmelt), a major cause of urban flooding and watercourse pollution. Within the City limits, Elbow river flooding has historically had a significant impact on houses, bridges, roads and other infrastructure. In the early 1900s, as construction of paved roads, sidewalks and houses increased, more and more stormwater was diverted directly into the Elbow. What sewer channels there were could no longer handle the volumes, so in 1927 the City built a storm sewer system. By the 1960s, the stormwater and wastewater systems were separated, and by the late 1990s, areas proposed for development were required to address stormwater management. Since then, the relevant municipalities have continued to develop policies to protect the rivers and reduce the impact of stormwater drainage.

One innovative way of dealing with smaller amounts of stormwater is the creation of constructed wetlands. These artificial wetlands are built as a mitigative measure, to reduce peak discharges and downstream flooding through detention storage, and also to remove pollutants from urban stormwater runoff through wetland uptake, retention and settling. In other words, they are built to do the job of the natural wetlands, which were previously removed to allow for development (a policy currently being reversed). Immediately east of Griffith Woods is the Elbow Valley Constructed Wetland — the second City focal point for environmental education within the watershed. This facility was built by the City of Calgary to deal with stormwater from the new subdivisions north and west of it; since 1999, it has been one of the busiest outdoor science classrooms in Calgary.

Nevertheless, floods along the Elbow have continued throughout the City's history. In 1929, a summer flood on the Elbow took out the bridge at 25th Avenue Southeast and created a lake in Victoria Park! Although

Wetlands: Worth Their Weight in Gold

Wetlands — one of the most productive ecosystems on earth — are areas that have water at, near or above the land surface, or land saturated with water long enough to support aquatic or wetland processes. In other words, at some time of year, these areas are wet. Their characteristics vary widely, as they can be ephemeral, temporary, seasonal, semi-permanent or permanent. No matter what type, they perform a critical function in our landscape. They improve water quality by filtration; provide habitat for waterfowl, reptiles, amphibians, fish and other species; act as reservoirs to reduce flooding; provide oxygen and water vapour to the air; and support recreational activities such as bird watching, hiking and fishing.

Wetlands cover about 20 percent of Alberta's surface area, but nearly all are north of the Elbow. Wetlands in southern Alberta are primarily temporary and semi-permanent ponds and marshes, and here is the bad news: nearly 70 percent of pre-settlement wetlands have disappeared, mainly filled in for agriculture. Even worse, in Calgary, 90 percent have been filled in for development, and the City estimates that 8,000 more wetlands could be impacted by city expansion. Responding to the enormity of such loss, the province has developed a new Wetland Policy and Calgary has adopted new guidelines for protecting their wetlands. Good stuff.

much urban damage was prevented by the newly excavated Glenmore reservoir in the infamous 1932 flood, the subdivisions along the Elbow experienced significant flood damage. Today, despite all previous mitigative measures, the communities in the Elbow's floodplain still experience the effects of the Elbow's flood waters. The worst flood on record ravaged the area in June 2013. The historic communities of Elbow Park, Rideau and Roxboro were once again inundated, as were Elboya, Mission, Stanley Park, Cliff Bungalow, Erlton, Victoria Park and Stampede Park. The 2013 floods were the costliest disaster in Canadian history in terms of insurable damage, and they forced government to deal with flood mitigation and management.

One of the best ways to experience the urban reach of the Elbow and view its history is from the Elbow River pathway, a system touted as early as 1933 by Parks Superintendant William Reader. It extends over 28 kilometres from the mouth of the Elbow up to and around the Glenmore Reservoir, part of the City of Calgary's 500-kilometre river-focussed pathway system. In fact, cycling upstream beside the Elbow from its mouth follows the expansion of the city over about 140 years and helps to demonstrate the part the Elbow has played in the City's history.

The influence of the Elbow began with pre-Calgary settlement history. By 1870, it was obvious that American traders were invading

The Challenge of Wastewater

Disposal of wastewater is perhaps the less appealing side of the municipal responsibility related to water. Calgary's original primitive sewer system, built before 1900, required continual expansion as the city grew. None of the wastewater was treated, however, based on the belief that dilution would deal with all the waste (per the common dictum: "the solution to pollution is dilution"); rather, it was dumped directly into the Bow River from a wastewater discharge point (outfall) near 1st Street Southeast. Aboveground, conditions were different then; the manure of the hundreds of horses stabled in, and moving through, the City fouled the streets. After a rain, the streets were awash in 15 centimetres of liquid manure and mud, which ultimately contaminated the groundwater and the rivers.

As wastewater from multiplying residential areas, industrial plants and the stockyards increased, those downstream of the downtown outfall cried foul. In 1914, after a period of intense public pressure, a trunk line was built to an outfall at Bonnybrook, on the southeast side of the City. Not until two decades later was the wastewater treated in a primitive primary waste treatment plant built there. Today, however, the situation is much different. The two wastewater treatment plants on the Bow are among the best in North America. The condition of the Bow River has gone from smelly and polluted, to once again a clear river supporting aquatic life in abundance. These wastewater issues have largely bypassed the fortunate Elbow so far.

Missions, Cowboys and an Urban Thirst

Canadian territory and more control was immediately required. In the Elbow watershed, Fred Kanouse, the infamous whisky trader euphemistically described by Dempsey as "the first merchant to locate at the future site of Calgary," operated his trading post on the Elbow, likely in the Britannia/River Park area. Sam Livingston had a trading post much farther up the Elbow at about the same time, although emphatically not dealing in firewater. Then, from Fort Macleod, the newly formed NWMP dispatched its 50-man F Troop to the Elbow-Bow confluence to lay down the law. Six weeks after arriving, by Christmas 1875, the Troop had floated pine and spruce logs down the Bow and Elbow rivers. With the help of men from the nearby I.G. Baker Company, The Elbow (or Bow) Fort was constructed as their rudimentary base of operations.

Canada's Wild West Settles Down (A Little)

European settlement and urban development had a slower start in this southern prairie area than around the early fur-trading forts to the north. Based partly on the Palliser Expedition's report that the southern prairies were unsuitable for agriculture, and partly on the negative statements of Sir George Simpson (Governor of the HBC), who wished to keep the area open for the fur trade, western settlement was not initially encouraged. But in 1869, the history of the West took a new turn. An all-Canadian railroad route across the country was determined feasible, and federal policy-makers turned to the importance of facilitating western settlement, if only to keep the land-hungry Americans at bay.

Fort Calgary, 1881, looking across the Elbow (foreground)
(Glenbow Archives NA-235-2)

This small palisaded fort, on a low rise between the two rivers, contained barracks and officers' quarters, stables and associated sheds, and a hospital, all with sod roofs and log walls chinked against the strong winter winds. (A replica currently stands near this site.) Briefly renamed Fort Brisebois by the unpopular troop commander, Captain Ephram Brisebois, it was soon redesignated Fort Calgary by Lt.-Col. James Macleod, the Commissioner of the NWMP, after his family estate Calgary House (from the Gaelic *Cala-ghearridh*, meaning "bay farm") on the Scottish Isle of Mull.

Almost immediately, the two major businesses operating in the region set up operations near the Fort. From the south gate in the log palisades, the Mounties could see the trading post of the I.G. Baker Company, a mercantile operation from Montana. Looking east across the Elbow, they could track the activity around the Hudson's Bay Company store, which sat squarely beside the trail leading south to Fort Macleod. By now 70 settlers scattered sparsely throughout the vicinity.

Roman Catholic Mission, Elbow River.

Our Lady of Peace Mission on the Elbow River, the first church in Calgary (Glenbow Archives NA-1434-40)

With the Mounties adding stability to the region, religious missions began congregating in the area of the Fort. Whether by happenstance or intent, the Catholic missionaries demonstrated an affinity for the Elbow River over the next century. The Oblates (the dominant Catholic missionary order in Canada's West and North) had established Our Lady of Peace (Notre-Dame-de-la-Paix) as the first permanent mission in southern Alberta, two years before the NWMP arrived. Father Constantine Scollen, an Irish priest with an affinity for languages, built this mission among the Blackfoot about 40 kilometres upstream on the Elbow. Hearing that the NWMP would shortly arrive in the area, he had a similar three-metre-square log building and an adjacent tent for added space put up at the junction of the Elbow and the Bow rivers. This was the first church in this settlement, ready and waiting when the Mounties arrived. The tiny building, however, was soon surrounded by police operations, so it was given to the NWMP and a larger one (a full six by seven metres in size!) built nearby. As Calgary was soon named an Oblate missionary district and headquarters for the Blackfoot missions, a third was built a few years later about two kilometres upstream on the Elbow, on the land where the Holy Cross Hospital later stood. A fourth building became the headquarters of the Oblate Mission (near 19th Avenue and 1st Street Southwest).

By 1882, the indomitable Father Albert Lacombe had arrived on the scene. A Metis from Lower Canada, he was preceded by stories of his previous missionary exploits among the Blackfoot and Woods Cree of the North-West Territories, where he earned the name, The Man of the Good Heart. Nearly two decades earlier, he had passed through the Elbow watershed, becoming friends with Sam Livingston and John Glen, the earliest settlers, before returning east. Father Lacombe had then become Superior of the Catholic Mission for all of what is now southern Alberta. This middle-aged energetic priest was focused on buying land for a permanent location for his Mission and for settlement of French-Canadians in Western Canada.

Teaching nuns and girl students outside the Sacred Heart Convent, late 1880s (Glenbow Archives NA-3981-11)

The assignment of much of the surveyed land to the railway company meant an impending rush on any available land close to the Bow-Elbow confluence. So, wasting no time, Father Lacombe began construction of a residence and chapel where the Sacred Heart Convent later stood in the Mission district. Next, this practical visionary claimed two homesteads (in the names of himself and Father Hippolyte Leduc) on the quarter-sections now occupied by the districts of Mission and Rideau-Roxboro. The priest had thus firmly established the basis for a French Catholic settlement south of the Fort and the railway line. The settlement was named Rouleauville, in honour of Chief Justice Charles B. Rouleau (Stipendiary Magistrate of the North-West Territories), and his brother, Hector (first Chief Surgeon of the NWMP).

Then began the creation of Catholic schools and hospitals west of the Elbow. In 1885, the Sisters of the Faithful Companions of Jesus (renowned teachers displaced by the North West Rebellion) were invited to open St. Mary's Girls School. In 1888, seeing a desperate need for a hospital, the Church contacted the experienced Grey Nuns of Montreal, who assembled nursing staff for a modest six-bed hospital built on 18th Avenue. In January 1891, in the middle of a frigid winter night, four Sisters arrived by train and doggedly made their way south to the Mission, carrying their own luggage. Three months later, they opened the little Holy Cross Hospital and were inundated with those needing their ministrations.

First Holy Cross Hospital, operated by the Grey Nuns, 1891 (Glenbow Archives ND-5-1)

By 1889, Rouleauville had grown to include 500 French-speaking families and St. Mary's Cathedral (the sixth iteration of the church in the Calgary area) was built, using sandstone from the quarry on the Butland farm by the Elbow River. Close to the growing city, the Grey Nuns soon found that their tiny hospital was woefully inadequate. After only a year of operation, it was replaced (in the first of many expansions) by a larger hospital housing 35 patients. Built beside the Elbow on the present Holy Cross property, with the

From Fort to Sandstone City to Banner City: Calgary by 1894

Following the destruction by fire of many of the wooden buildings in downtown Calgary in 1886 and again in 1889, sandstone was decreed the construction material for large buildings. In addition to the Butland quarry on the Elbow, the 14 other sandstone quarries operating beside the Bow and Elbow rivers provided an ample supply from the underlying Paskapoo formation, to the extent that Calgary soon became known as the Sandstone City.

By 1894, Calgary seemed like a real metropolis on the prairie: electric street lights, a prospering rebuilt commercial downtown, a daily newspaper and daily rail service to Edmonton (taking hours instead of days by stagecoach). Musical entertainment was offered by four bands and performers at the Opera House. The Calgary Brewing and Malting Company produced fine ales for consumption. The town communicated with far-flung settlements not only by telegraph, but by telephone as well. The venerable Ranchmen's Club (the oldest continuing institution in Calgary today) provided a gentleman's club for well-to-do ranchers who needed a place to stay and socialize in town. So with a population of nearly 4,000 people serving the ranching, farming and railway service industries, Calgary was successfully incorporated as a city. It was not long until the city began promoting itself as the Banner City of the Last Great West. (Rather like Calgary's more recent Heart of the New West marketing approach.)

---

substantial improvements of hot and cold running water, baths and electricity, the Holy Cross Hospital was operated by the Nuns for nearly eighty years. In 1969 it was sold to the provincial government, which had assumed responsibility for health care, and closed in 1996. Until the 2013 flood, it operated privately as the Holy Cross Centre, a medical facility providing general and specialist clinics and learning centres. Only the façade of the McNab Wing and the historic name remain of the original.

At the turn of the century, Calgary was still a small town, surrounded by open prairie, with milk cows and horses kept in many back yards. Travellers to Calgary on the Macleod Trail often saved a couple of rutted kilometres by veering left onto the Mission Road detour, to cross the Elbow River on the narrow iron Mission bridge. The Blue Rock (later Albion) Hotel on the north bank signalled the entrance to Rouleauville, and a sign on the hotel said One Mile to Calgary. This hotel, a popular watering hole, engendered a rather unique use of the Elbow River: inebriated cowboys were rolled down the bank into the water to sober up.

By 1915, the Mission area was filling up with houses on narrow eight-metre lots, and with small retail stores and services on Broadway (4th Street). Then, as now, the cool tree-lined valley and broad Elbow River provided welcome respite from the noise and bustle of the surrounding city. South of the Elbow, the Lacombe-homesteaded land had been sold by the Church. This was prime land — situated by the winding river, accessible to the downtown by street car, close to schools and a hospital — and developers jumped on it. One big-thinking developer, Fred C. Lowes, decided that the Roxboro area was too low. Using high pressure hoses, he pumped Elbow River water at the steep river bank to cut away soil and wash it into the Roxboro area. (If he had added a few more metres, this area might have escaped the floods of 1932 and 2013!) Then, having transformed the floodplain, he built a new-concept subdivision, with trees, sidewalks, graded streets and electric lights. As the lots were immediately snapped up, similar development took place in the Rideau subdivision, then Elbow Park, Glencoe and Elboya, all the way south up the Elbow to Windsor Park.

The Elbow valley has hosted recreational activities since the early settlement of Calgary. Calgary's first hockey game was played on the Elbow in 1888, and its first fall horse racing meet was held in the Elbow valley in 1890. As the wealthy and the growing middle classes had more time for recreational pursuits, winter festivals were organized, including horse and buggy racing on the river. A favourite family activity was skating on the Elbow. In fact, as archivist John Gilpin describes, until the Glenmore Dam was built in 1933, skaters would glide all the way from Weaselhead Flats to the lower Elbow. Several swimming holes, including two of the best-used at Elbow Park (Lowes)

New Century, New Province, New Enterprise

Early in the century there was a critical housing shortage. Pierre Berton claimed "the demand for space exceeded the supply by ten to one: thousands lived in tents, barns, and shacks," most of them near the Elbow and Bow rivers. As has been the case ever since, Calgary's solution was to annex additional lands (like Rouleauville, which became the Mission district) and create new subdivisions in concert with land developers. The value of land skyrocketed, reaching a zenith in 1912, when there were more real estate offices in the city than grocery stores.

In 1905 Alberta became a province, and Calgary was bitterly disappointed not to be chosen as the capital city. Two years later, Edmonton was again chosen over Calgary, this time for the location of the province's first university. However, the great entrepreneurs of Calgary regarded these as minor setbacks, and carried on with their energetic enterprise, full steam ahead. Even the severe winter of 1906–07 and the influx of farming homesteaders, both of which seriously impacted the open-range cattle industry, did not slow construction in the city.

and at 24th Avenue, were provided with booms, change rooms and lifeguards.

"Ice Matinee" horse race on the Elbow River, 1906 (Glenbow Archives NA-1451-34)

For the snow-free months, a group of dedicated golfers decided that they required a proper course to play on. Since 1897, the Calgary Golf Club had been playing on scrub land south of the CPR tracks; by 1906, they had moved up the Elbow to Elbow Park, but even this location proved inadequate. Two years later, they secured a favourable area further south in a horseshoe bend of the Elbow, and this became the Calgary Golf and Country Club. This tony club occupies the same prime riverside real estate today, 116 years later, and has one of the longest waiting lists for membership in Canada. It was the first of many, as there are nearly a dozen golf courses in the Elbow watershed today.

On the west side of the Elbow River in another of its horseshoe bends, about one and a half kilometres upriver from the Fort, another line of recreation developed early, one based on the major agricultural and business initiatives of the Calgary area. In 1888, Calgary's Agricultural Society bought 38 hectares of land from the federal government, named it Victoria Park, and created a home for an annual agricultural fair, with a modest exhibition building, sheds for cattle and a race track for the increasingly popular horse races. In 1908, additional buildings, a roofed grandstand and barns were added as Calgary proudly hosted the federally funded Dominion

Exhibition on this site. Soon a new livestock and horse show arena was under construction, and pari-mutuel betting was introduced at the thoroughbred and harness races. These races were held there for 117 years, until Calgary Stampede expansion plans forced an unpopular decision in 2008: no more horse racing in Stampede Park.

Outside the city, the ranching and cowboy culture of the area spawned small informal rodeos farther up in the Elbow watershed in Springbank, Bragg Creek and other locations, as farmers and ranchers gathered for friendly competition and neighbourly socializing. The skills involved — riding bucking horses, roping calves, riding bareback — were part of ranch and farm life, but the competition was exhilarating. Venues like the Robinsons' Elbow Park Ranch provided such exhibitions as entertainment for their constant flow of guests, royal and otherwise. And building on a Tsuu T'ina tradition, in 1896 an official rodeo was held in Moccasin Flats on the Elbow River, in present-day Mission.

In 1912, Calgary's future was recast when American vaudeville Wild West performer Guy Weadick came to town with a plan for a frontier show. Four prominent businessmen-ranchers (the Big Four) provided financial support so that the history of the big open-grazing ranches and the skills of the "true cowboy" would be recognized and remembered. The five-day show held in September 1912 was a resounding success, attracting 80,000 people (when the city's

Cowboy Clem:
World Champion and Community Mainstay

Clem Gardner, who took over the Gardner ranch and eventually owned one of the largest cattle herds in the west (2,000 head), loved to compete in rodeos and races, especially the Calgary Stampede. In 1912 he won third place for steer-roping, tied for third in bucking-horse riding, and was selected as the best all-around Canadian cowboy. He competed in calf-roping until he was 60 years old. For many years, Clem supplied the livestock for the Stampede, trailing hundreds of long-horned cattle, brahma bulls and bucking horses into town and then back to his ranch at the end of the rodeo. Clem promoted the change to lighter chuckwagons and thoroughbred horses, competed every year between 1923 and 1946, and proudly won the world championship in 1931. ➔

Clem Gardner riding "High Tower" at the Calgary Stampede, 1919 (Glenbow Archives NA-590-1)

→ He had other accomplishments too. Following in his horseman father's footsteps, he played on local and international polo teams. In 1957, Clem completed the deeding of land on the Elbow River for a Scout camp called Camp Gardner, which is still operational today. In recognition of his exemplary involvement in his community, the 1955 Highway 22 bridge over the Elbow was named in his honour.

population was only 60,000) and gave birth to the legendary Calgary Stampede. Assembled in locations like the Belgian Horse Ranch in Springbank, stock for the Stampede were driven into town, sharing the routes with buggies and automobiles, and finally fording the Elbow River to reach the Exhibition stables. Local cowboys and ranchers, including Stanley Fullerton from Bragg Creek and Elbow Valley rancher, Clem Gardner, son of Meopham, were enthusiastic participants in the rodeo events.

Chuckwagon race on a fast and dusty track in the 1920s
(Glenbow Archives NA-1644-25)

In 1923, the Stampede merged its exciting ranch-oriented entertainment with the agricultural tradition of the Calgary Industrial Exhibition to become the Calgary Exhibition and Stampede, billed worldwide today as

The Greatest Outdoor Show on Earth. In that year, Weadick introduced chuckwagon races, the opening Stampede Parade, sidewalk breakfasts, storefront decorating, and "dressing western," all of which remain Stampede attractions today. After 1923, the Stampede became an annual event and for its centenary in 2012, the federal government recognized it as a National Historic Event, for preserving both an historic western Canadian way of life and the distinctive cultures of First Nations participants.

From the outset, southern Alberta First Nations peoples, including the Tsuu T'ina, played a large part in the Stampede — in the Parade, the Indian Village and rodeo competitions. The Indian Village has provided a rare opportunity for the over one thousand Treaty 7 participants to share their traditional culture with visitors from around the world. Despite this goodwill and cooperation, it was not until 1992 that a First Nations representative became part of Stampede management. Roy Whitney, then Chief of the Tsuu T'ina Nation, was finally appointed to the prestigious Board of Directors, and two months later the Nation's Bruce Starlight was elected.

Tipis in the colourful Indian Village at the Calgary Stampede, 2010

The Stampede is much more than a rodeo. The agricultural exhibition, one of the most unique in the world, has a thousand exhibitors providing displays and demonstrations ranging from blacksmith competitions to stock dog trials. Also from the beginning, famous American western painter Charles Russell and others have showcased the role of art in creating an image of the West.

The Stampede grounds, sitting solidly within a generous curve of the Elbow, have expanded significantly since 1912, and now include a legacy of the 1988 Olympic Games, a hockey and large event arena (the distinctively-shaped Saddledome which hosts the National Hockey League's Calgary Flames). The Calgary Stampede today is a mammoth not-for-profit organization supported by an army of 2,000 volunteers and 1,200 full-time staff, with an annual budget of about $130 million. Providing entertainment for over a million visitors in 10 days each July, Stampede Park also hosts over a thousand additional events every year.

The Elbow cradles this hive of western-based activity within its sinuous flow, but their relationship has not always been friendly. In the 1950s, after several floods, the Stampede lobbied for room to expand. The city considered one proposal — to divert the Elbow into a concrete canal, removing one of the meanders which would then be backfilled — but abandoned it in favour of other flood control measures above the Dam. In June 2013, these measures were not enough, as Stampede Park was completely inundated by floodwaters, which filled the Saddledome up to its eighth row of seats. With less than two weeks until the opening of the 101st Stampede, and with *Hell or High Water* as their slogan, the organization mounted a monumental, and successful, effort to pump out the water and clean up the debris. The Elbow continues to meander through this reach, but now there is a more serious plan for flood mitigation, including a possible diversionary tunnel from the Glenmore Dam out to the Bow River.

Throughout the hundred years of Stampede Park operations, the Elbow river banks and water quality have been subjected to serious degradation. Finally, the Stampede has installed storm water interceptors under their parking lots to divert contaminants from the river, and their 2009 Master Plan for Stampede Park includes development of additional green space and "beautification" of the banks of the Elbow. As part of an overall revitalization of this area (part of the new "Rivers" area in east Calgary), a planned Riverwalk on the west bank of the Elbow will connect Stampede Park with Fort Calgary, Inglewood and a Bow River walkway to Chinatown. This will enhance access to, and appreciation of, the Elbow by the general public throughout the year.

So much activity, so much history, so great a concentration of population in this narrowest part of the Elbow watershed. The watershed has shrunk from its 40-kilometre width near the river's source in the mountains, to five kilometres wide in this reach, finally decreasing to only a few metres at its mouth. Throughout this final urban stretch, the Elbow winds its independent way past houses and apartment buildings, golf courses and dog parks, streets and pathways, shops and medical clinics, horse barns and hockey rinks, down to its termination. This is its meeting point with the larger Bow, in a grassy field not far from the steel and glass towers of downtown Calgary and the historic fort where the city had its beginnings. This is the destiny of every river, to merge itself with a larger one until together they reach the sea. With attention to good management, a river should reach that confluence as healthy as it was at its source. It can be done.

Part 4.

Whither the Elbow?

Chapter 11.
Reflections

"We have to protect our water. Because in the end, what we do to water, we do to ourselves."
—**ROBERT SANDFORD**, *WATER AND OUR WAY OF LIFE*

Pristine Elbow Lake in the upper watershed (photo courtesy of Bryan Mercer)

I count myself a most fortunate being. Alongside my good friends and family, I have explored most of a watershed. I have wandered around in the headwaters of the Elbow River, in the dramatic landscape of the Front Ranges, and attempted to understand where those mountains came from and why they look as they do today. I have sat in the watershed's sensitive alpine meadows, marvelling at the tiny pink campion and the curious climate-challenged pikas. I have enjoyed the amenities of a beautiful mountain backcountry campground, sitting beside a clear headwater lake under the stars. For the moment, it seems almost pristine.

Then, walking downriver into the watershed's subalpine zone, I have pondered the hydrologic cycle which largely governs the quantity of water available to the watershed, and have thought about the people who have been in this elevated place – the mountain adventurers and map-makers who put names to the magnificent features of the landscape. How did they tread on this landscape? Very lightly.

Having the freedom to backpack through the relatively unspoiled area where the mountains meet the foothills is a privilege. Here the watershed is so broad and so varied I cannot get to all of it. But to sit by the Elbow and examine its river ecosystems — riparian, benthic and aquatic — is an education. Finding diverse subalpine flora, each with special traits and uses — like valerian, cow parsnip and the larch tree — makes every hike unique.

Lower in the watershed the traffic begins — the zone of large campgrounds and a two-lane paved highway which runs up the valley from Bragg Creek carrying cars, horse trailers, and many many RVs. There are more hikers, bikers and equestrians on the trails, and evidence everywhere of industrial activity. But, you know what? It is still a calm, lovely and endlessly interesting place to be. The big foothills ridges are inviting and so accessible. What I have learned about butterflies and climate change, periglacial features and potholes, and the power of running water in forming river channel patterns and waterfalls continues to amaze me. Rivers, waterfalls and powerful landforms inspire us, as so many artists attest.

In the foothills industrial development zone of this watershed, what Moose and Prairie mountains have meant to people over time seems currently overshadowed by the impact of oil and gas operations, forest companies and cattle grazing. "Pristine" has disappeared and I am sorry to see the land increasingly criss-crossed with pipeline and seismic rights-of-way, forest clearcuts and roads. Are we doing our best to protect this vulnerable region of our watershed? I believe we could and should do better. This can start with education, providing a better understanding of the forests themselves — their succession and the impacts of fire, wind and insects — and the consequences of their removal from the watershed in an unsustainable way.

The Elbow River snaking through the foothills

In this foothills region, the effects of various types and qualities of land management are much more apparent than in the mountains. Perhaps the busiest stretch of the foothills watershed runs from Allen Bill to Bragg Creek, where recreationists abound in all seasons of the year. Camping, OHVing, fishing, canoeing, kayaking, hiking, biking, RVing — all these activities attract participants from the local area, from Calgary, from other parts of the province and from elsewhere in Canada and the world. The pressure builds, affecting biodiversity in this zone, pushing wildlife further up in the watershed, changing the composition of plant communities. Here good management of the land and water is key, to enhance the recreation experience, and to protect the environment from overuse by recreationists and unsustainable use through industrial activities.

In this river reach is found the hamlet of Bragg Creek, straddling the Elbow. The quality of water in its groundwater and in the river downstream is an issue, yes, but it is now being dealt with. Meanwhile, its dedication to its history, its culture and its natural environment is inspiring, as is its resilience when confronted with difficult times such as the 2013 flood.

Below Bragg Creek in the lumpy aspen parkland zone, extending all the way to the meandering Elbow's mouth, everything is different again. Here, the history of the people of the watershed is fascinating — native peoples, settlers and the thread of horse history that ties it all together. What is perhaps most salient is the meaning of the river to the people of this watershed — people who lived off the land here, who explored here, who homesteaded here, who finally settled here. I have always found the ongoing relationship between the landscape and the people within it particularly engrossing: each has impacted the other, and not always with respect. Does understanding their historical perspective make it even more challenging to evaluate how we use the land and water today, as conversion of agricultural land to houses and commercial properties continues apace, as we recognize a polluted aquifer around the Elbow, and as a shortage of water looms? Sometimes we long for the good old days, although they were neither easy nor always that good!

Then, a slight digression to an unusual watershed focus — the military. The depth of military history within this little watershed is surprising, a contribution out of proportion with the rest of the surrounding region and city. Beginning in a small way, the Elbow watershed actively supported the military for training, armed conflicts and peacekeeping, and today honours its proud history through its museums, geoglyphs, parks and even commemorative housing developments.

The Elbow River from Calgary's Mission Bridge

The last reach of the river is the one within the City of Calgary, where the Elbow River is celebrated for its contribution to slaking the urban thirst, and enjoyed through the riverside parks and extensive pathway system. It is also censured for the disastrous floods it imposes on its floodplain. Indeed, the history of the city is mirrored by a trip up the Elbow, from its mouth and Fort Calgary, through Rouleauville-Mission, to riverside subdivisions, and the century-old Calgary Stampede.

There it is: the Elbow watershed. Wandering through the watershed in this journey, we now know much more about its history and its present. What about its future? What will it look like in fifty or a hundred years? That will depend on what each of us does today.

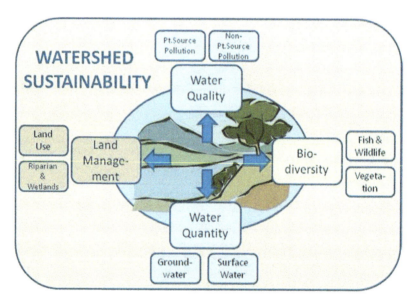

Diagram of the critical components of watershed sustainability

Maybe we can start by agreeing that our watershed is an ecological unit that functions best when all its parts are healthy. This is key. Looking at the elements of the watershed — the climate, the physical landforms, the geology, the soil, the vegetation, the fish and wildlife, the people and their activities, the infrastructure, and most importantly, the river and its tributaries, riparian areas, wetlands and groundwater — we have a sense of how all these parts are interconnected and interdependent. We also better understand the forces that act upon this special watershed, environmental and human. The environmental forces — floods, hail, drought, fire, pests, wind — will continue. In some cases, we humans can modify our negative impacts, but we must do this in a sustainable way.

Then we should probably consider the water itself. Water quantity and water quality have always been inextricably linked. The amount of drinking water available, for example, ultimately depends on the quality of the water, whether in the source water or via the treatment plant's purification. Predicted climate change means warming temperatures, lower and/or less predictable precipitation, extreme weather events, longer and more intense droughts, increased evapotranspiration, retreating glaciers, decreasing mountain snowpack, and reduced river flows and groundwater recharge. All of these impact both the water supply and water quality, just as the demand for water is growing.

Less water in the river means that the natural function of the river in filtering out contaminants from stormwater, agricultural runoff and industrial effluents is reduced, and the concentration of pollutants is increased. Less water in the river also means that water temperature rises; warmer water plus more concentrated pollutants lead to excessive plant growth through eutrophication, the most prevalent global water quality problem. Less water in the river and its aquifer adversely affect not only the supply of drinking water and the cost of providing it both publicly and privately (as on rural properties), but also fish and wildlife habitats, recreational activities, and irrigation and stock watering. Many scientists and water managers now understand that land use and water policies need to be considered together, not separately as in the past. Water quantity and water quality must be managed as one.

The human forces acting on the watershed are complex. Official management of the watershed is in the hands of several authorities: Alberta Environment and Sustainable Resource Development, Kananaskis Improvement District, Rocky View County, the City of Calgary, the Tsuu T'ina Nation. Organizations with input to watershed management include watershed stewardship organizations like the Elbow River Watershed Partnership and the Bow River Basin Council, and environmental groups like Cows and Fish and the Weaselhead Preservation Society. Industry has a very significant presence in the Elbow watershed, whether it be oil and gas companies like Shell Canada and Husky Energy, or the FMA operator, Spray Lake Sawmills, or ranchers with cattle grazing permits, or the urban developers. Not least, the individual landowners, like me, greatly affect the watershed's health collectively through our rural and urban land use and water management practices.

With such complex governance, planners and legislators must work together on proactive land and water management policies which will deal with the watershed as an integrated whole. Protecting source waters and bringing ecological stability into the discussion have become more important in recent years − a marked change from prior approaches. But the stumbling block is the actual implementation of all that good stuff. Who takes responsibility and who enforces the regulations? Who has the resources?

We all can do a better job of managing the watershed. For example, municipal development permitting should require small building footprints, use of permeable hard surfaces, retention of significant amounts of native green space within and around each development and inclusion of stormwater management systems. Every property should have a hookup to an approved water supply and wastewater treatment facilities (instead of rural septic systems) plus rain collection and recycling. Riparian setbacks can be further increased to protect these critical zones, and construction over the aquifer strictly prohibited. Rather than spending money on additional infrastructure in light of a declining water supply, reducing the demand through water conservation and efficiency — making the most of the water available — seems a much more sustainable municipal option. To increase the supply of water, planners should more frequently consider rainwater harvesting, use alternative water sources (such as greywater for some uses) and water recycling (wastewater for agriculture).

Good management cannot happen without good information. One area which needs much more detailed mapping and analysis is the groundwater in the Elbow watershed, and the aquifer under the river in particular. A start has been made, but the political will must be there to complete the job, and then to continue the monitoring and the mitigative measures.

Industry needs further regulation and enforcement. For example, the regulatory goal for forestry, oil and gas, agricultural, and urban development operations should be protection of the integrity and stability of natural ecosystems within the watershed. Industrial management plans and field operations will therefore have to be modified, and regulatory authorities will have to monitor the results.

And all stakeholders need to focus on watershed education. As Grant MacEwan said, "Change begins with a sense of stewardship which itself grows out of understanding." Educational initiatives require support to raise public awareness about the ultimate value of water as a non-renewable resource. Over time, those stakeholders will understand the new reality. Reinforcing this premise, David Suzuki (in noted science writer Fred Pearce's *When the Rivers Run Dry)* has clearly stated, "We have to re-examine our relationship with water, learning from ancient insights that it is not simply a

"Unfortunately, in Alberta water literacy is low and few people understand the complexity of water-based ecosystems. Successful watershed management requires many groups, with many different interests, working together to ensure the resource is sustainable for multiple uses."

—Sydney Sharpe, *Alberta: A State of Mind*

commodity for commerce but a sacred substance that maintains all life on the planet." Fred Pearce emphatically agrees that central to the public education process is the awareness that not just our own, but humanity's future on this planet is entirely dependent on the fate of our rivers. Some individuals and institutions have already been motivated to change their thinking about water as a critical resource, and in the way they use water. (For example, per capita water consumption for single family homes in Calgary has been reduced from 325 litres per day in 2003 to 231 litres per day in 2013.) But so far, too few.

Never too young to get to know the Elbow

It really all comes down to the individual and what we do on a daily basis. We need to arm ourselves with water knowledge and do what we can to reduce the amount of water and energy we use, and the way in which we use them, through relatively simple and cost-effective means. Our actions do matter.

We can pass on our knowledge to help educate others about watershed concerns, and those of the Elbow in particular, by talking to our friends and neighbours.

Reflections

We can contact those in authority about our watershed concerns, such as inappropriate development over the Elbow's aquifer or destruction of riparian habitat.

We can change the small things we do to conserve water and to maintain or improve its quality. (For example, in *Water, Weather and the Mountain West*, Robert Sandford provides 70 practical ideas for "What you can do to save water.")

We can join organizations who care about and work on behalf of the Elbow's watershed, like the ERWP or the Friends of Kananaskis, and make our vision known to them.

We can volunteer for projects that improve the quality of the land and water in the watershed, like bioengineering in McLean Creek PLUZ to remediate stream crossings.

We can attend forums or workshops given by government or industry to educate ourselves and to provide our comments.

We can honour the watershed by writing a piece of music or a poem about the song of the river, or by painting a beautiful picture of the watershed, with whatever our particular talent might be.

We can take our children and grandchildren down to the river and to the Elbow Valley Constructed Wetland and for a hike in the upper watershed.

And we can look into what is happening in another watershed nearby or one of special interest to us.

"Watersheds...[are]... the perfect instruments for understanding the totality of the human experience of place."

—Robert Sandford, *Water and Our Way of Life*

And we cannot give up, even if it sometimes seems our watershed is losing the battle.

Our Elbow watershed is itself at a watershed. We can make it happy and healthy if we are persistent. As we know now, throwing even one stone into a river will change something in that river. So pick up your stone and get started. Make ripples and they will turn into waves.

Appendices

Appendix 1.

Historical Timeline: Elbow River Watershed and Region

CALGARY AND REGION	DATES	ELBOW RIVER WATERSHED
Continental and cordilleran glaciation, further erosion and deposition, and creation of our present Rocky Mountains, foothills and plains landscapes	1.6 million years ago	Meeting of continental and cordilleran ice sheets in the Elbow watershed area
Habitation in the region by pre-Clovis people, likely camping close to present-day Vermilion Lakes near Banff	12,000 years ago	Glacial Lake Calgary receded, Bow and Elbow valleys formed
Most of Alberta free of ice; extinction of the ice-age mammals; formation of present ecological zones	10,000 years ago	Paleo-Indians settled in watershed/Calgary area, hunting bison
Worldwide glacial expansion as part of Little Ice Age Hudson's Bay Company received Charter to trade in Rupert's Land (western Canada) (1670)	1600s	Tsuu T'ina regularly moving through Elbow watershed
David Thompson explorations in Alberta (1787)	1700s	Tsuu T'ina acquired horses (after 1750) Fort La Jonquiere established at mouth of Elbow River (1751-1785)

CALGARY AND REGION	DATES	ELBOW RIVER WATERSHED
Tsuu T'ina became part of the Blackfoot confederacy, territory focussed on Red Deer and Bow rivers	1800–1850	Peigan nation controlled the land of the Elbow watershed (by 1815)
Canadian Confederation, formed the Dominion of Canada (1867)	1850–1870	Mount Rae, Mount Head, Fisher Range named by Palliser (1859); Palliser called the Elbow River "Swift Creek" on his map and followed it westward
Dominion Lands Act (1872), encouraged settlement of western Canada	1870s	Fred Kanouse, Sam Livingston built trading posts on Elbow (1871, 1873) Notre Dame de la Paix Mission at Elbow River crossing (1873) Fort Calgary built by NWMP at Bow/Elbow confluence (1875)
Bison exterminated (1889) Dominion grazing leases encourage ranching in southern Alberta (1881) Canadian Pacific Railway line reached Calgary (1883)	1880s	Oblate missionary Father Lacombe arrived (1882); Rouleauville established Tsuu T'ina given reserve in the Elbow and Fish Creek watersheds (1883) Chipman Ranche established (1883), Bridge built across Elbow in Mission (1887) for city access from south Cheese factories built (Springbank in 1888 (first in Alberta); Robinson in 1889)
North West Irrigation Act passed, reserving all water rights in name of the Crown (1894) Alberta water allocation based on First in Time, First in Right (FITFIR) principle (1894)	1890s	Coal prospecting (Dr. George Ings), later mining below Moose Mountain (1910s) Holy Cross Hospital opened in Mission district (1892) Copithorne ranch enterprise started (1893) Albert and John Bragg homesteaded in Bragg Creek area (1894) Dominion Surveyor A.O. Wheeler surveyed much of Elbow Watershed, part of foothills survey for irrigation planning (report published 1896) Pirmez Irrigation Society (south side of Elbow) (oldest irrigation society in Alberta) (1895)

Historical Timeline: Elbow River Watershed and Region

CALGARY AND REGION	DATES	ELBOW RIVER WATERSHED
Calgary "unprivatized" the water supply and redesigned water system (1900) Alberta became a Province (1905) Dominion Exhibition held in Calgary (1908) Street car system (Calgary Municipal Railway) (1909) built in Calgary Original survey carried out for Forest Reserve west of Calgary (1909), including Elbow Watershed	1900–1910	Lord Strathcona's Horse formed (1900), sent to South Africa (1901) for Boer War First water license on Elbow River granted to Meopham Gardner (1902) Water flume (wooden stave pipeline) built to take water from Elbow River to Calgary (1907) (operated until 1934) Calgary annexed village of Rouleauville, became Mission district (1907)
End of "Great Ranches" era (~1910)	1910s	First Calgary Stampede held in Victoria Park (1912) Calgary subdivisions built in Glencoe, Elbow Park, Roxboro, Rideau, Elboya and Windsor Park (1912) First bridge across the Elbow at Bragg Creek built (1913) Mowbray-Berkeley oil well drilled near Elbow River (1913) Sarcee Training Camp set up (1914) 50th Battalion, Canadian Expeditionary Force, formed (1914), trained at Sarcee Camp, sent to France (1915) Signal Hill battalion numbers placed (1915-16) Elbow Ranger Station built (1915) Stampedes re-established, starting with "Victory Stampede" (1919)

CALGARY AND REGION	DATES	ELBOW RIVER WATERSHED
Alberta horse population peaked (more horses than people) (1920) CFAC and CFRN radio stations begin broadcasting in Calgary (1922) Imperial Oil opened refinery in Calgary (1923) Seebe/Sarcee power line built by Calgary Power, running through the Springbank area (1924) Dingman well produced oil in Turner Valley, starting three-year boom (1924) Calgary Aero Club (later Calgary Flying Club) formed (1927) (by 1929, was the largest flying club in Canada) Calgary built storm sewer system, separate from the sewer system (1927).	1920s	Chuckwagon races added to the Calgary Exhibition and Stampede (1923) Upper Elbow store built in Bragg Creek (now the Trading Post) (1925) Five oil wells drilled in Elbow valley (1927, 1928, 1935) Moose Dome Oils sunk two wells (1927, 1935) several miles up Canyon Creek, fuelled by own natural gas Calgary Aero Club's, and Calgary's, first airfield opened on Banff Coach Road (1928) Herron Petroleum Co. drilled well between Ranger Station and Elbow Falls (1928) Elbow Oil Company drilled well near Hwy. 66/758 junction (1928); shut down in 1929 Major flood on Elbow River (1929) Moose Mountain fire lookout built (1929)

Historical Timeline: Elbow River Watershed and Region

CALGARY AND REGION	DATES	ELBOW RIVER WATERSHED
Natural Resources Transfer Act gave Alberta control of its natural resources (1930) Provincial Parks and Protected Areas Act (1931) passed in Alberta Alberta Water Resources Act: all water the property of provincial government (1931) Bonnybrook Wastewater Treatment Plant was the first sewage primary treatment facility for Calgary (~1935).	1930s	Camp Cadicasu founded by RC Diocese of Calgary (1930) Herron Petroleum drilled a well near the Elbow Ranger Station (1930's) Dawson Hill School built (1931) on Bragg Creek road; closed in 1943 and moved to Bragg Creek (now an ECS kindergarten) Two Pines School built (1931) to serve south Bragg Creek Work camps in Bragg Creek area cut fire guard on Forest Reserve boundary (1931) Mount Royal became a Junior College, affiliated with the University of Alberta (1931) Glenmore reservoir land acquired; Tsuu T'ina sold 240 ha of their reserve to the City (1931) Major flood on Elbow River (1932) First youth hostel in North America built at Elbow/Bragg Creek confluence (1933) Currie Barracks opened to house military personnel (1933), for Lord Strathcona's Horse Glenmore Dam completed and reservoir flooded (1933) Springbank and Elbow Valley set up mutual telephone companies (1936, until 1969) Construction of roads and runways at Currie Barracks (1936-39) and Currie Field (1935) Pirmez Creek Irrigation Society formed (used same ditches as Calgary Irrigation Company in 1893) (1939) Two RCAF squadrons based at Currie Barracks (1939)

CALGARY AND REGION	DATES	ELBOW RIVER WATERSHED
Population of Calgary at 89,000 (1940) Calgary Brewing and Malting fish hatchery largest game fish hatchery in Canada (1942) Last log drive down the Bow River to Calgary (1943), ending a 56-year practice Establishment of the provincial Green, Yellow and White Zones (1948)	1940s	No. 3 Service Flying Training School (SFTS) opened on Currie Field (1940 until 1945) Currie Barracks an A-16 infantry training centre (1940) Avro Anson training aircraft crashed on Mount McDougall (1941) Army log bridge built across Elbow River (Hwy 22) (1943) Springbank Rural Electrification Association brought power to rural areas (first co-operative in Alberta) (1947) Currie Field renamed RCAF Station Lincoln Park (NATO pilot training centre) (1947–1958) Mayview Golf Club (later Pinebrook) began operation west of city (1947)
Over half of Alberta's population was urban (1951) Indian Act amended to make ceremonial practices legal (1951) City of Calgary western boundary moved to 69th Street (1956)	1950s	Kamp Kiwanis founded by Kiwanis Club (1951) Fire lookout built on Forgetmenot Mountain (1952) Tsuu T'ina leased reserve land to Department of National Defence (1958) Lincoln Field used only as emergency landing facility (1958-1964) Clem Gardner Bridge replaced log bridge on Hwy 22 road (1955) Camp Gardner (Scouts) given land on Elbow River (1957) Hail suppression equipment installed on Moose Mountain (1957)
Calgary's second wastewater treatment plant (Fish Creek) began operation (1960) Calgary's storm and sewage wastewater systems fully separated (1960)	1960s	Bragg Creek Provincial Park established (1960) Heritage Park opened beside Glenmore Reservoir (1964) RCAF Station Lincoln Park closed (1964) Easter Seals Camp Horizon built by Kinsman Club (1965) Currie Barracks redesignated as CFB Calgary (1966) "Springbank Park for All Seasons" construction began (1968)

Historical Timeline: Elbow River Watershed and Region

CALGARY AND REGION	DATES	ELBOW RIVER WATERSHED
Bow River extremely polluted below the City of Calgary (1970) Bonnybrook Wastewater Treatment Plant upgraded to provide secondary wastewater treatment (1971) Calgary's Centennial celebrated (1975) Policy for resource management on the Eastern Slopes of Alberta produced (1977), citing watershed protection as a top priority	1970s	Springbank Airport opened (1971) Mount Royal College opened Lincoln Park campus (1972) on old Lincoln Field Springbank Community High School built (1975) Elbow River Estates acreage community developed (1976) (first rural residential development in area) Redwood Meadows Golf Club established near Bragg Creek on Tsuu T'ina land (1976)
Calgary installed tertiary treatment facilities at the Bonnybrook and Fish Creek Wastewater Plants (early 1980s) Calgary's Light Rail Transit system began operation (1981) Winter Olympic Games held in Calgary (1988)	1980s	Highway 66 built in the Elbow valley Sarcee Barracks redesignated as Harvey Barracks, before being returned to Tsuu T'ina (1981) Calaway Park (Western Canada's largest outdoor amusement park) opened (1982) McLean Pond and Allan Bill Pond dammed on Elbow River (1983) — later Forgetmenot Pond as well Kananaskis Country established (1983) Shell Oil built pumping station on Moose Mountain, pumping gas to Esso's Quirk Creek Plant (1985) DND/Tsuu T'ina agreement for cleanup of previously leased lands (1985)
Canadian Pacific Railway head office moved to Calgary (1996) City of Calgary implemented largest ultraviolet wastewater treatment system in the world (1996)	1990s	Part of Sarcee Training Area land (1500 ha) returned to Tsuu T'ina (1990) Battalion Park (Signal Hill) created and designated as historic site (1991) Elbow-Sheep Wildland Provincial Park established (1996) Holy Cross Hospital closed (1996) CFB Calgary decommissioned and developed for housing (1998)

CALGARY AND REGION	DATES	ELBOW RIVER WATERSHED
Alberta government issued *Water for Life* strategic planning document (2003) (updated 2008) Calgary Exhibition and Stampede Centennial celebrated (2012) Serious flooding in southern Alberta, in Bow, Elbow, Highwood and other watersheds – costliest disaster in Canadian history (2013)	2000–present	Griffith Woods Natural Area opened (2000) Harvey Barracks land returned to Tsuu T'ina Nation (2000) Don Getty Wildland Provincial Park established (2001) Large Elbow Valley residential development begun west of City of Calgary (early 2000s) Formation of Elbow River Watershed Partnership (ERWP) (2002) Major Elbow River floods (2005, 2013) Moratorium on new water licenses include Elbow River (2006) Mount Royal College became Mount Royal University (2009) Tsuu T'ina built casino near Elbow River in Calgary (2007); expansion began in 2012 Tsuu T'ina approved land deal with City of Calgary for western ring road (2013)

Appendix 2.

Scientific Names

alder *(Alnus* spp.*)*
alpine buttercup *(Ranunculus eschscholyzii)*
alpine forgetmenot *(Myosotis alpestris)*
alpine larch *(Larix lyallii)*
antelope (pronghorn) *(Antilocapra americana)*
Apollo butterfly *(Parnassius smintheus)*
aspen poplar *(Populus tremuloides)*
balsam poplar *(Populus balsamea)*
bearberry *(Arctostaphylos uva-ursi)*
beaver *(Castor canadensis)*
bighorn sheep *(Ovis canadensis)*
bison *(Bison bison)*
black bear *(Ursus americanus)*
black-backed woodpecker *(Picoides arcticus)*
blackfly *(Simulium* spp.*)*
blue grouse *(Dendragapus obscurus)*
blueberry *(Vaccinium* spp.*)*
boreal chickadee *(Parus hudsonicus)*
bracted honeysuckle *(Lonicera involucrata)*
brook trout *(Salvelinus fontinalis)*

brown trout *(Salmo trutta)*
brush wolf *(Canis latrans)*
buffalo beans *(Thermopsis rhombifolia)*
bull trout *(Salvelinus confluentus)*
caddisfly (Order Trichoptera)
Canada buffaloberry *(Shepherdia canadensis)*
chipmunk *(Eutamias* spp.*)*
Clark's nutcracker *(Nucifraga columbiana*
cougar *(Felis concolor)*,
cow parsnip *(Hercaleum lanatum)*
coyote *(Canis latrans)*
crowberry *(Empetrum nigrum)*
cutthroat trout *(Onchorhynchus clarki)*
daisy fleabane *(Erigeron* spp.*)*
damselfly (Order Zygoptera)
dawn horse *(Eohippus)*
Dolly Varden trout *(Salvelinus malma)*
dragonfly (Suborder Anisoptera)
dwarf birch *(Betula glandulosa)*
elephant-head *(Pedicularis groenlandica)*
elk *(Cervus elaphus)*
Engelmann spruce *(Picea engelmannii)*
fireweed *(Epilobium angustifolium)*
golden eagle *(Aquila chrysaetos)*
golden-mantled ground squirrel *(Spermophilus lateralis)*
grizzly bear *(Ursus arctos)*
grouseberry *(Vaccinium scoparium)*
hare *(Lepus* spp.*)*
harebell *(Campanula* spp.*)*
harlequin duck *(Histrionicus histrionicus)*
heart-leaved arnica *(Arnica cordifolia)*
hoary marmot *(Marmota caligata)*
honeysuckle *(Lonicera involucrata)*
horse *(Equus caballus)*
Hungarian partridge *(Perdix perdix)*

juniper *(Juniperus* spp.*)*
kingfisher *(Ceryle alcyon)*
Labrador tea *(Ledum groenlandicum)*
lodgepole pine *(Pinus contorta)*
lynx *(Lynx canadensis)*
mammoth *(Mammuthus primigenius)*
mayfly (Order Ephemeroptera)
moose *(Alces alces)*
moss campion *(Silene acaulis)*
mountain avens *(Dryas drummondii)*
mountain pine beetle *(Dendroctonus ponderosae)*
mountain goat *(Oreamnos americanus)*
mountain sheep *(Ovis canadensis)*
mule deer *(Odocoileus hemionus)*
nettle *(Urtica gracilis)*
paintbrush *(Castilleja* spp.)
paper birch *(Betula papyrifera)*
pika *(Ochotona princeps)*
pine grosbeak *(Pinicola enucleator)*
pink wintergreen *(Pyrola asarifolia)*
porcupine *(Erethizon dorsatum)*
prairie crocus *(Anemone patens)*
prairie wolf *(Canis latrans)*
prickly rose *(Rosa acicularis)*
purple finch *(Carpodacus purpureas)*
rainbow trout *(Onchorhynchus mykiss)*
raspberry *(Rubus* spp.)
red paintbrush *(Castilleja miniata)*
red squirrel *(Tamiasciurus hudsonicus)*
Richardson's water vole *(Arvicola richardsoni)*
Russian thistle *(Salsola australis)*
river beauty *(Epilobium latifolium)*
rock jasmine *Androsace chamaejasme)*
roseroot *(Sedum rosea)*
ruby-crowned kinglet *(Regulus calendula)*

ruffed grouse *(Bonasa umbellus)*
Saskatoon berry *(Amelanchier alnifolia)*
shooting star *(Dodecatheon* spp.*)*
short-tailed weasel *(Mustela erminea)*
silky scorpionweed *(Phacelia sericea)*
snipe *(Gallinago gallinago)*
snowshoe hare *(Lepus americanus)*
spruce budworm *(Choristoneura* spp.*)*
spruce grouse *(Dendragapus canadensis)*
star-flowered Solomon's seal *(Smilacina stellata)*
stonecrop *(Sedum* spp.*)*
stonefly (Order Plecoptera)
subalpine fir *(Abies lasiocarpa)*
sweetvetch *(Hedysarum* spp.*)*
tent caterpillar *(Malacosoma* spp.*)*
trembling aspen *(Populus tremuloides)*
twinflower *(Linnaea borealis)*
valerian *(Valeriana* spp.*)*
warbler *(Dendroica* spp., *Wilsonia pusilla)*
water shrew *(Sorex palustris)*
water vole *(Microtus richardsoni)*
white mountain aven *(Dryas octopetala)*
white spruce *(Picea glauca)*
white-tailed deer *(Odocoileus virginianus)*
wild rose *(Rosa acicularis)*
willow *(Salix* spp.*)*
woodpecker *(Picoides* spp.*)*
wolf *(Canis lupus)*
wolf willow *(Eleagnus commutata)*
yellow columbine *(Aquilegia flavescens)*
yellow mountain avens *(Dryas drummondii)*

References Cited:

Alberta Government. 2003. Water for Life: Alberta's Strategy for Sustainability. Edmonton. 31p. Accessed 24 March 2009 at environment.gov.ab.ca.

—. 2007. Mountain Pine Beetle Action Plan. Accessed 2 March 2009 at http://mpb.alberta.ca/AlbertasStrategy/documents/MPB_action_plan.pdf.

—. 2008. Water for Life: A Renewal. Edmonton. 18p. Accessed 24 March 2009 at http://environment.gov.ab.ca/info/posting.asp?assetid=8035&searchtype=advanced&title=water%20for%20life:%20a%20renewal.

—. 2009. Water for Life: Action Plan. Edmonton. 24p. Accessed 22 November 2013 at http://environment.gov.ab.ca/info/library/8236.pdf.

—. 2013a. Draft South Saskatchewan Regional Plan: 2014-2024. Edmonton. 157p.

Alberta Riparian Habitat Management Society. 2013. Riparian Health Summary Final Report: Elbow River Watershed 2012

Riparian Health Inventory Project, Kananaskis Country Alberta. Cows and Fish, Airdrie AB. 28p.

Alberta Tourism, Parks and Recreation. 2008. Draft Management Plan for Provincial Recreation Areas–Kananaskis Management Area. Edmonton. 47p.

Alberta Water Council. 2012. Review of Implementation Progress of Water for Life: 2009–2011. Edmonton. 41p. Accessed 26 November 2013 at www.albertawatercouncil.ca.

Allen, J. A. and J. L. Carr. 1947. Geology of the Highwood-Elbow Area, Alberta. Research Council of Alberta Report No. 47. King's Printer, Edmonton. 75p. + appendices.

Barlow, M. and T. Clarke. 2002. Blue Gold: The Battle Against Corporate Theft of the World's Water. Stoddart Publishing Co., Toronto. 278p.

Berton, Pierre. 1984. The Promised Land: Settling the West 1896-1914. McClelland and Stewart Limited, Toronto. 378p.

—. 2001. Marching as to War: Canada's Turbulent Years 1899-1953. Doubleday Canada, Toronto. 610p.

City of Calgary. 2014. Calgary Watershed Report: A Summary of Surface Water Quality in the Bow and Elbow Watersheds, 2010-2012. City of Calgary. Pdf document. 93p.

Conaty, G.T. 2004. The Bow: Living With a River. Glenbow Museum, Calgary AB. 159p.

Daffern, Gillean. 1997. Kananaskis Country Trail Guide. Volume 2. 3rd Edition. Rocky Mountain Books, Calgary AB. 320p.

—. 2011. Kananaskis Country Trail Guide, Volume 2: West Bragg, The Elbow, The Jumpingpound, 4th Edition. Rocky Mountain Books, Calgary AB. 320p.

Dempsey, Hugh A. 1994. Calgary: Spirit of the West. Glenbow and Fifth House Publishers, Calgary AB. 159p.

—. 1997. Indian Tribes of Alberta. Glenbow Museum, Calgary AB. 108p.

Doll, J. et al. 2003. Elbow River watershed assessment: What has changed in the past 14 years? Slide presentation, accessed at www.ucalgary.ca/ensc/files/ENSC502 02-03 Elbow River Watershed.pdf on 21 June 2013. Environmental Science, University of Calgary, Calgary. 164p.

Eastern Slopes Grizzly Bear Project. 1998. Grizzly Bear Population and Habitat Status in Kananaskis Country, Alberta: A Report to the Department of Environmental Protection, Natural Resources Service, Alberta. Prepared by the Eastern Slopes Grizzly Bear Project, University of Calgary, Calgary, Alberta. Accessed 17 March 2009 at http://www.canadianrockies.net/grizzly/rspub.html.

Foothills Historical Society. 1976. Chaps and Chinooks: A History West of Calgary. Volume 1 and Volume 2 (1900-1945). 703p. Foothills Historical Society, Calgary AB. Also accessed at http://www.ourroots.ca/e/toc.aspx?id=4048.

Gadd, Ben. 1995. Handbook of the Canadian Rockies (2nd Edition). Corax Press, Jasper AB. 831p.

Gilpin, John. 2010. The Elbow: A River in the Life of the City. Detselig Enterprises, Calgary AB. 263p.

Herrero, Stephen (editor). 2005. Biology, demography, ecology and management of grizzly bears in and around Banff National Park and Kananaskis Country: The final report of the Eastern Slopes Grizzly Bear Project. Faculty of Environmental Design, University of Calgary, AB. 276p. pdf accessed 14 April 2013 at http://www.canadianrockies.net/grizzly/final_report.html.

Leopold, A. 1949. A Sand County Almanac, and Sketches Here and There. Oxford University Press, New York NY. 226p.

MacEwan, Grant. 2000. Watershed: Reflections on Water. NeWest Press, Grant MacEwan College, Edmonton. 193p.

Pearce, Fred. 2006. When the Rivers Run Dry: Journeys into the Heart of the World's Water Crisis. Key Porter Books, Toronto ON. 400p.

Potter, Mike. 2008. Fire Lookout Hikes in the Canadian Rockies (2nd Edition). Luminous Compositions, Turner Valley AB. 272p.

Roland, J. and S. Matter. 2007. Encroaching forests decouple alpine butterfly population dynamics. PNAS 104:34, 3 p. August 21, 2007. Accessed January 2009 at www.pnas.org/cgi/reprint/0705511104v1.pdf?ck=nck.

Sandford, R.W. 2003. Water and Our Way of Life. The Rockies Network, Fernie BC. 160p.

—. 2007. Water, Weather and the Mountain West. Rocky Mountain Books, Surrey BC. 198p.

Sharpe, Sydney. 2005. Alberta: A State of Mind. Sydney Sharpe, Roger Gibbons, Heather Bala Edwards, James Marsh (Eds.). Key Porter Books, Toronto. 288p.

Sosiak, A. and J. Dixon. 2004. Impacts on Water Quality in the Upper Elbow River. Pub. No. T/740, Alberta Environment/City of Calgary, Calgary AB.

Trout Unlimited. 2003. Turbidity Monitoring in Howard Creek, Elbow River Watershed. Calgary AB. Accessed 26 March 2009 at www.erwp.org.

Web Sites Referenced in Text:

http://www.spraylakesawmills.com/Woodlands/ForestManagementAgreement/tabid/137/Default.aspx

http://www.tucanada.org/NP_QuirkCreek.shtml

http://www.wsc.ec.gc.ca/applications/H2O/index-eng.cfm

www.erwp.org

www.geo.ucalgary.ca/DIY/nihahi5

www.pc.gc.ca/cp-nr/release_e.asp

Maps Referenced in Text:

Municipal District of Rocky View. 1968. Land Ownership Map. Calgary AB.

Municipal District of Rocky View. 1997. Land Ownership Map. Calgary AB.

Rocky View County. 2009. Land Ownership Map. Calgary, AB.

Rocky View County. 2011. Land Ownership Map. Calgary, AB.

For More Information:

Alberta Environment. 2013. Alberta Wetland Policy (online pdf version) accessed 27 January 2014 from http://www.waterforlife.alberta.ca/documents/Alberta_Wetland_Policy.pdf.

Alberta Forest Service. 2005. The Alberta Forest Service: Protection and Management of Alberta's Forests. Chapter 1: Early Days. Edmonton AB. Accessed 9 April 2009 at www.esrd.alberta.ca/forests/researcheducation/afs.

Alberta Sustainable Resource Development. 2008. Alberta Grizzly Bear Recovery Plan: 2008-2013. Alberta Species at Risk Recovery Plan No. 15. Edmonton. 78p. Accessed at http://esrd.alberta.ca/fish-wildlife/wildlife-management/bear-management/grizzly-bears/documents/GrizzlyBear-RecoveryPlan2008-13-revJuly23-2008.pdf.

— and Alberta Conservation Association. 2010. Status of the Grizzly Bear (*Ursus arctos*) in Alberta: Update 2010. Alberta Wildlife Status Report No. 37 (Update 2010). Edmonton. 44p. Pdf document accessed 4 April 2013 at http://esrd.alberta.ca/fish-wildlife/species-at-risk/

species-at-risk-publications-web-resources/mammals/documents/SAR-StatusGrizzlyBearAlbertaUpdate2010-Feb2010.pdf.

Alberta Watersmart. 2013. The 2013 Great Alberta Flood: Actions to Mitigate, Manage and Control Future Floods. Watersmart Solutions Ltd., Calgary AB. Pdf document. 27p.

Alberta Wilderness Association. 2005. Fire in Alberta's Boreal Forests: Position Statement. Edmonton AB. Accessed 10 March 2009 at www.albertawilderness.ca.

Andriashek, L. and D. McKenna. 2008. Groundwater management in Alberta: Yesterday, today and what the future holds. Presentation to Alberta Environment Conference 2008. Accessed 6 October 2009 at http://www.environmentconference.alberta.ca/docs/Session-31_presentation.pdf.

Armstrong, C. and H. V. Nelles. 2007. The Painted Valley: Artists Along Alberta's Bow River, 1845-2000. Univ. of Calgary Press, Calgary. 160p.

Armstrong, C., M. Evenden and H. V. Nelles. 2009. The River Returns: An Environmental History of the Bow River. McGill-Queen's University Press, Montreal. 475p.

Bagley, F. and H. D. Duncan. 1993. Legacy of Courage: Calgary's Own 137th Overseas Battalion, C.E.F. Plug Street Books, Calgary. 252p.

Bow River Basin Council. 2005. Nurture, Renew, Protect: A Report on the State of the Bow River Basin. BRBC, Cochrane AB. 202p. Accessed at www.brbc.ab.ca/pdfs.

—. 2010. Bow River Basin State of the Watershed Summary 2010 (Web-based). BRBC, Calgary AB. Pdf document, 44p. Accessed at http://www.brbc.ab.ca/index.php/resources/publications/our-publications.

Caldwell, Mark. 1997. The wired butterfly: the world's tiniest radar tags are making a Rocky Mountain butterfly—and its ecology—easier to follow—the Apollo butterfly. Discover Magazine, February 1997. Accessed January 2009 at http://discovermagazine.com/1997/feb/thewiredbutterfl1053.

Calgary Herald. 2013. The Flood of 2013: A Summer of Angry Rivers in Southern Alberta. Greystone Books, Toronto. 136p.

Carter, Sarah. 1999. Aboriginal People and Colonizers of Western Canada to 1900. University of Toronto Press, Toronto ON. 186p.

Chen, Z. et al. 2006. Historical climate and stream flow trends and future water demand in the Calgary region, Canada. Water Science and Technology 53(10):1-11.

Copeland, Kathy and Craig Copeland. 2002. Where Locals Hike in the Canadian Rockies: The Premier Trails in Kananaskis Country near Canmore and Calgary. Voice in the Wilderness Press (www.wild.bc.ca). 246p.

Crain, E.R. 2000. The True History of Oil and Gas. Spectrum 2000. Canadian Well Logging Society. 4p. Accessed on 5 March 2009 at www.spec2000.net/freepubs/TrueHistory.pdf.

Dempsey, Hugh A. 2002. Firewater: The Impact of the Whiskey Trade on the Blackfoot Nation. Fifth House Ltd., Calgary AB. 243p.

Dickason, O.P. 2009. Canada's First Nations: A History of Founding Peoples from Earliest Times. With D.T. McNab. Oxford University Press, Don Mills ON. 574p.

Dyke, A. S. 2004. An outline of North American deglaciation with emphasis on central and northern Canada, *in* Quaternary Glaciations- Extent and Chronology, Part II, p. 373-424, J. Ehlers and P. L. Gibbard, eds, Elsevier B.V.

Elbow River Watershed Partnership. 2009. Final Report (revised January 2009): Elbow River Basin Water Management Plan: A Decision Support Tool for the Protection of Water Quality in the Elbow River Basin. 18p. + appendices. Accessed 12 February 2011 at www.erwp.org.

Elofson, Warren M. 2000. Cowboys, Gentlemen and Cattle Thieves: Ranching on the Western Frontier. McGill-Queen's University Press, Montreal. 196p.

Environment Canada. n.d. Pikas cannot beat the heat. Column 194: Your Yukon. Accessed at www.taiga.net/yourYukon/col194.html.

Fish and Wildlife Historical Society. 2005. Fish, Fur and Feathers: Fish and Wildlife Conservation in Alberta, 1905-2005. Fish and Wildlife Historical Society and Federation of Alberta Naturalists, Edmonton AB. 418p.

Geist, V. 1999. Mule Deer Country. NorthWord Press, Minnitonka MN. 175p.

Godwin, Ted. 1991. The Lower Bow: A Celebration of Wilderness, Art and Fishing. Hard Art Moving and Storage Co. Ltd., Calgary AB. 63p.

Jenish, D'arcy. 2004. Epic Wanderer: David Thompson and the Mapping of the Canadian West. Anchor Canada, Toronto ON. 300p.

Johnson, D.W. 1999. Biogeography of Quaking (Trembling) Aspen (*Populus tremuloides*). Biogeography 319 course notes, San Franscisco State University. Accessed 15 October 2009 at http://bss.sfsu.edu/geog/bholzman/courses/Fall99Projects/aspen.htm.

Lawby, C., D. Smith, C. Laroque and M. Brugman. 1995. Glaciological studies at Rae Glacier, Canadian Rocky Mountains. Physical Geography 15(5):425-441.

Leslie, Jean. 1994. Glimpses of Calgary Past. Detselig Enterprises Ltd., Calgary AB. 128p.

MacEwan, Grant. 1975. Calgary Cavalcade: From Fort to Fortune. Western Producer Book Service, Saskatoon SK. 200p.

Millarville Historical Society. 1979. Foothills Echoes. Millarville Historical Society, Millarville AB. Accessed at http://www.ourroots.ca/e/toc.aspx?id=7611.

Millarville, Kew, Priddis and Bragg Creek Historical Society. 1975. Our Foothills. Calgary AB. Accessed at http://www.ourroots.ca/e/toc.aspx?id=4134.

Mussieux, R. and M. Nelson. 1998. A Traveller's Guide to Geological Wonders in Alberta. The Provincial Museum of Alberta. Edmonton AB. 254p.

Newson, A.C. and D. Sanderson. 2000. Exploration Targets in the Rocky Mountain Foothills: Calgary to Moose Mountain—A

Helicopter Supported Field Trip. GeoCanada 2000, Calgary AB. Accessed at http://www.mooseoils.com/guidebook.pdf.

Ommanney, C.S. L. 2002. Glaciers of the Canadian Rockies, USGS Prof. Paper 1386-J-1, in, Satellite Image Atlas of Glaciers of the World, R.S. Williams Jr./J. G. Ferrigno (Eds.), J199-J289. Accessed 27 April 2011 at http://pubs.usgs.gov/pp/p1386j/canadianrockies/canrock-lores.pdf.

Potter, Mike. 2001. Ridgewalks in the Canadian Rockies. Luminous Compositions, Turner Valley AB. 304p.

Rasporich, A. and H.C. Klassen. 1975. Frontier Calgary: Town, City and Region 1875-1914. University of Calgary, McClelland and Stewart West, Calgary AB. 306p. Accessed at http://www.ourroots.ca.

Read, Tracey. 1983. Acres and Empires: A History of the Municipal District of Rocky View No. 44. Tall-Taylor Publishing, Irricana AB. 390p. Accessed at http://www.ourroots.ca.

Rollins, J. 2004. Caves of the Canadian Rockies and Columbia Mountains. Rocky Mountain Books, Surrey B.C. 336p.

Rosenberg International Forum on Water Policy. 2007. Report of the Rosenberg International Forum on Water Policy to the Ministry of Environment, Province of Alberta. Univ. of California, Berkeley CA. 26p. Accessed 14 November 2007 at www.erwp.org.

Rutherford, S. 2004. Groundwater Use in Canada. West Coast Environmental Law. 30p. Accessed 21 August 2009 at http://wcel.org/resources/publication/groundwater-use-canada.

Ryan, C. 2008. Alberta's alluvial aquifers and rivers: Truly a single resource. Power Point presentation to Alberta Wilderness Association Headwaters Workshop, November 6, 2008, Cochrane AB. Accessed at http://albertawilderness.ca/issues/wildwater/headwaters/headwaters-archive/20081106_CRyan.pdf/at_download/file.

Schindler, D.W. and W.F. Donohue. 2006. An impending water crisis in Canada's western Prairie Provinces. Proc. National Academy of Sciences. USA 109(19):7210-7216. Accessed 4 April 2011 at http://www.ncbi.nlm.nih.gov/pmc/articles/PMC1564278/.

Strategic Relations Inc. (SRI). 2013. Alberta's Feral Horses: Managing Populations. Consultation Report, Alberta Environment and Sustainable Resource Development, Edmonton. 17p. Accessed 1 February 2014 at www.esrd.alberta.ca/lands-forests/land-management/feral-horses/documents/AltaFeralHorses-Managing Populations - Apr 26-2013.pdf.

Sturgess, P.K. 2014. Elbow River Historical Detention and Diversion Sites. Alberta Watersmart report to Alberta Flood Recovery Task Force, Government of Alberta, Edmonton. 19p.

Vander Ploeg, Casey. 2010. From H2O: Turning Alberta's Water Headache to Opportunity. Report for Alberta Water Research Institute. Canada West Foundation. 120p. Pdf accessed 9 March 2011 at http://cwf.ca/CustomContentRetrieve.aspx?ID=1108674.

Volney, W.J.A. and J.R. Spence. 2003. Managing boreal forest insect disturbances for sustainability. XII World Forestry Congress, Quebec City QC. Accessed 16 March 2009 at www.fao.org/DOCREP/ARTICLE/WFC/XII/0789-B1.

Glossary

Anastomosing river: a river with more than one thread (channel) which splits around floodplain elements; each channel can have independent flow and sediment patterns

Anticline: an arch of stratified rock in which the layers bend downward in opposite directions from the crest, often trapping hydrocarbons in the arch

Aquifer: a geologic unit of permeable rock or loose material that can store and transmit water; a water-bearing formation

Arête: a sharp, narrow mountain ridge, often between two gorges

Banff formation: a stratigraphical unit of Devonian age overlying the Palliser formation in the Canadian Rocky Mountains, consisting of siltstone, shale, sandstone, chert and marl

Banks, right and left: river banks identified based on their orientation as the viewer looks downstream

Boreal forests: forests of cold-hardy trees, such as spruce, fir and pine

Brachiopod: a mollusk-like marine invertebrate of the phylum *Brachiopoda*, with two hinged shells (lamp shell)

Braided stream: one with slightly curved, multiple, shallow and wide channels that repeatedly branch out and reunite, with numerous islands resulting

Calcareous: containing calcium carbonate, chalky

Cap rock: a hard or resistant rock or rock layer that sits above less resistant material or an oil- or gas-bearing formation

Catchment area: the drainage area that contributes water to a particular point along the channel (i.e., everything upstream of the sampling point)

Clear-cut: a harvesting and regeneration method that removes all trees within a given area; clear-cutting is most commonly used in pine and hardwood forests

Climax forest: a forest community that represents the mature stage of natural forest succession for its environment

Cobble: a rock fragment between 64 and 256 millimeters in diameter, especially one that has been naturally rounded

Col: a low pass or saddle between peaks, on a ridge or between drainage systems in a watershed

Glossary

Cretaceous period: the last period of the Mesozoic era, from 145 to 66 million years ago

Devonian Palliser formation: a stratigraphical unit found in the Main and Front Ranges of the Rocky Mountains, consisting of fossil-bearing dark limestone in the upper part and massive dolomitic limestone in the lower part

Discharge: water flowing into and within a river

Dolostone: a sedimentary rock composed mainly of the mineral dolomite

Dry dam: a flood-retarding structure, without turbines or gates, that allows the water to flow freely except in times of intense flow, when it holds back water that would otherwise cause flooding downstream

Ecozone: a part of the environment that has similar geography, vegetation and animal life; a biogeographic realm

Eutrophication: the process by which water bodies receive nutrients from runoff, leading to excessive plant growth and depletion of oxygen

Feral animal: an animal in a wild state after escape from captivity, or one born to such an animal in the wild

Firefinder: a directional device (alidade) used over a map to find a directional bearing (azimuth) for detected wildfire smoke

Fossil: a remnant or impression of plants and animals typically found in stratified rock layers, mainly from species extinct today

Freshet:	a river flood resulting from heavy rain or melted snow
Greywater:	wastewater from domestic activities, such as laundry and dishwashing, which can be recycled for some uses
Groundwater:	water that is found everywhere under the ground, and flows through water-bearing formations (aquifers)
Headward erosion:	the fluvial process that lengthens a stream opposite to the direction of flow, and thus enlarges its drainage basin, by eroding soil and rock at its headwaters
Ice age:	a glacial episode during a past geological period
Inlier:	an area of older rocks surrounded by younger rocks
Interstitial ice:	the ice formed in the narrow gaps between small rocks and/or soil sediment
Invertebrate:	an animal lacking a backbone, such as an arthropod, mollusk, annelid or coelenterate
Limestone:	a layered sedimentary rock consisting mainly of calcium carbonate
Little Ice Age:	a 300-year cold period said to last from 1550-1850 C.E.
Lower Carboniferous Rundle limestone:	the Lower Carboniferous period spanned 345 to 310 million years ago; the Rundle Group is mostly thick, erosion-resistant limestone and dolomite, with layers of shale, siltstone and sandstone
Macroinvertebrates:	insects without backbones

Glossary

Meandering stream:	one with a single, sinuous, deep and narrow channel and few islands
Mesozoic era:	the interval of geologic time between 252 and 66 million years ago, following the Paleozoic era; also called the Age of Reptiles
Moraine:	an accumulation of glacial debris (soil and rock) built by the direct action of glacial ice, usually in the form of a long crested ridge
Outfall:	the discharge point of a waste stream, river or drain where it empties into a body of water
Paleoclimate:	the climate of some former period of geologic time
Paleozoic era:	the interval of geologic time between 541 and 252 million years ago, preceding the Mesozoic era, divided into six geologic periods (Cambrian, Ordovician, Silurian, Devonian, Carboniferous and Permian); also called the Era of Ancient Life
Paleozoic Mississippian:	a subperiod of the Paleozoic Carboniferous period, between 358 to 323 million years ago
Pekisko formation:	a limestone formation, part of the Rundle Group, of Mississippian age, found in the foothills and plains of Alberta
Periglacial:	of or relating to near-glacial environments
Photosynthesis:	the process by which green plants and some other organisms use sunlight to synthesize foods from carbon dioxide and water

Pleistocene epoch:	the geological epoch which lasted from 2.6 million to 11.7 thousand years ago, including the most recent periods of frequent glaciations; a subdivision of the Quaternary Period (2.6 million years ago to present); often called the Age of Humans
Plunge pool:	a pothole occurring at the foot of a waterfall
Provincial significance:	areas which either have features with limited distribution in Alberta or which provide the best examples of a feature in Alberta
Reach:	a length of river channel which has relatively uniform characteristics (e.g., slope, depth, discharge)
Recessional moraine:	the end moraine formed during a temporary standstill of ice while the glacier is receding
Sandstone:	a sedimentary rock formed by the consolidation and compaction of sand-sized particles (mineral, rock, organic material) and held together by a natural cement, such as silica; one of most common types of sedimentary rock
Scree:	a loose collection of stones and rock fragments forming a steep slope on the side of a mountain (also called talus)
Sensitive species:	species which are not currently endangered but which require active management or conservation to prevent them from becoming at risk

Shale:	a fine-grained sedimentary rock formed from cemented clay and silt
Special Protection Natural Park:	one with a high level of biodiversity and wildlife which is accorded the highest level of protection and management.
Subalpine zone:	the zone on middle mountain slopes between the highest-growing aspen and the treeline, in which tree growth becomes progressively stunted due to exposure and heavy snow cover
Talus:	rock fragments derived from and accumulated at the base of a steep slope or cliff (also called scree)
Unconfined aquifer:	an aquifer with no confining or impermeable upper limit; its upper surface is the water table, the location of which depends on the amount of water in the aquifer
Vulnerable species:	a species likely to become endangered, unless the conditions of its survival and reproduction improve
Wastewater:	water that has been adversely affected in quality by human activity, as in washing, manufacturing, flushing or agricultural operations

Index

Symbols

2nd Canadian Mounted Rifles (CMR) 158
103rd Regiment (Calgary Rifles) 158

A

Alberta Environment and Sustainable Resource Development 82, 201, 237
Alberta Fish and Wildlife 39, 94, 97, 108
Alberta Forest Service 14, 60, 231
Alberta Riparian Habitat Management Society 95, 221
Alberta Stewardship Network 92, 97
Alberta Water Council 93, 222
Allen Bill 76, 95, 98, 100, 102, 105, 110, 123, 198
alluvium 78
alpine xi, 8, 9, 10, 11, 12, 13, 14, 16, 20, 22, 23, 29, 40, 43, 54, 55, 58, 80, 104, 196, 217, 224
anastomosing river 239
anticline 239
Apollo butterfly 54, 217, 233
aquatic zone 38

aquifer	126, 239
arête	81, 239
aspen	xi, 17, 21, 66, 67, 84, 87, 99, 104, 114, 140, 141, 142, 143, 174, 175, 198, 217, 220, 235, 245

B

Banded Group	27, 31, 32, 33, 34, 35, 43, 46, 61
Banded Peak	31, 35, 46, 120
Barclay, Catherine and Mary	118
Barlow, Maude	89
Battalion Park	162, 215
beaver	47, 67, 141, 174, 175, 217
beaver bafflers	175
Beaver Flat Campground	67
Beaver Flats	175
Beaverlodge Trail	175
Belgian Horse Ranch	148, 188
benthic zone	38, 40, 47
Berton, Pierre	163, 185
Big Elbow River	43, 44, 46
bioengineering	92, 93, 205
bison	94, 134, 135, 136, 137, 209, 217
Blache, Louis Napoleon	148, 157
Boer War	158, 211
boreal	xi, 16, 21, 22, 52, 99, 100, 140, 217, 237
Bow River	xi, 18, 52, 70, 71, 105, 114, 124, 127, 137, 158, 178, 190, 191, 202, 215, 232, 233
Bow River Basin Council	92
brachiopod	55, 59, 240
Bragg, Albert Warren	112
Bragg Creek	xii, 37, 47, 75, 76, 77, 88, 89, 90, 102, 106, 111, 112, 113, 115, 116, 118, 119, 120, 121, 123, 124, 126, 127, 128, 129, 130, 134, 137, 138, 140, 146, 147, 148, 166, 167, 187, 188, 196, 198, 210, 211, 212, 213, 214, 215, 235

Bragg Creek Provincial Park 111
braided stream 240
Burgess, Thornton W. 145
Burgie Ranch 147, 148, 150

C

Calgary Flames 190
Calgary Golf and Country Club 186
Calgary Highlanders 166, 168
Calgary Irrigation Company 148, 213
Calgary Ring Road 168
Calgary Stampede 46, 167, 187, 188, 190, 200, 211, 212, 216
Camp Cadicasu 116, 213
Camp Gardner 116, 148, 188, 214
Camp Horizon 98, 116, 214
Canada Lands Company 169
Canadian Forces Base Calgary 159
Canadian Forces Base (CFB) Calgary 167
Canadian Pacific Railroad 142
Canyon Creek 55, 74, 75, 76, 79, 82, 94, 102, 108, 118, 212
Canyon Creek Ice Cave 79
cap rock 68
catchment area 240
Chief Bull Head 137
Chipman Ranche 142, 146, 210
cirque 8, 9, 10, 29, 33, 40
City of Calgary xii, 17, 96, 123, 126, 127, 140, 148, 154, 159, 175, 177, 178, 200, 202, 215, 216, 222, 225
clear-cut 75, 88, 89, 96, 116, 240
climate change 47, 196, 201
climax forest 240
coal 13, 18, 52, 55, 75, 76, 77, 108, 114
cobble 240

Cobble Flats	63, 102
col	240
Conaty, Gerald	70
constructed wetlands	177
Cooper Memorial Hall	151
Copithorne, Richard and John	148, 151, 210
Cougar Mountain	27, 45, 46
Cows and Fish	95, 202, 222
Cox, Richard	158
coyote	xiii, 37, 84, 95, 99, 114, 144, 145, 148, 174, 218
Create-a-Park campaign	115
Currie Barracks	162, 164, 165, 166, 167, 213, 214
Currie, Sir Arthur	163

D

Daffern, Gillian	23, 29, 33, 222
dawn horse Eohippus	134
Dawson, George Mercer	4, 5, 27
Dawson Hill School	151, 213
Dempsey, Hugh	136, 140, 160, 179, 223, 233
Desolation Flat	17, 27, 28
Dominion Exhibition	187
Dominion Lands Act	136, 141, 210
Don Getty Wildland Provincial Park	58, 216
Drummond, Major Patrick (Paddy)	158
dry dam	65, 124, 241

E

ecozone	241
Edworthy Falls	68
Elbow Falls	63, 67, 68, 72, 94, 98, 102, 104, 105, 110, 212
Elbow Lake	4, 5, 9, 13, 14, 17, 19, 24, 44, 116, 135
Elbow Loop	31, 32, 44, 46, 104

Elbow Park	146, 147, 158, 178, 185, 186, 211
Elbow Park Ranch	146, 147, 158, 187
Elbow Pass	3, 24, 82
Elbow Ranger Station	76, 78, 83, 105, 211, 213
Elbow River Estates	143, 215
Elbow River pathway	178
Elbow River Provincial Park	89, 110
Elbow River Rifle Club	162, 164
Elbow River Watershed Partnership	92, 93, 95, 96, 97, 110, 202, 205, 216, 234
Elbow-Sheep Wildland Provincial Park	14, 215
Elbow Valley Hall	151
Elbow Valley Trail	108, 143
Elpoca mountain	7, 18, 22, 44
Elpoca Mountain	7, 17
Elsdon, Jack	125
eutrophication	201, 241

F

felsenmeer	59
Fidler, Peter	136
firefinder	84, 241
fire lookout	60, 74, 75, 80, 82, 83, 109, 212
First in Time, First in Right (FITFIR)	154, 210
First Nations	viii, xiii, 14, 18, 71, 81, 92, 99, 125, 136, 137, 140, 145, 189, 234
forest fire	60, 66, 80, 81, 82, 84, 107
Forest Management Agreement	87, 88, 89, 202
Forest Reserve	76, 82, 105, 107, 114, 122, 149, 213
forest succession	84, 240
Forgetmenot Pond	63, 74, 215
Forgetmenot Pot	59
Forgetmenot Ridge	52, 57, 59, 60
Fort Brisebois	180

Fort Calgary	70, 112, 116, 137, 142, 157, 180, 191, 200, 210
fossil	34, 55, 58, 59, 241
Fraser, J.A.W.	148
freshet	125, 127, 242
Front Ranges	4, 7, 21, 31, 33, 46, 47, 52, 57, 72, 100, 124, 196, 241
Fullerton, Jake	76, 116, 118, 123, 129
Fullerton Loop	117
Fullerton, Stanley	188
Fullerton, Thomas Kerr (T.K.)	116, 142, 157

G

Gardner, Captain Meopham	147, 151, 157, 211
Gardner Cattle Company	151
Gardner, Clem	187, 188, 214
geoglyphs	160, 161, 199
Gilpin, John	185
Glenbow Museum	70, 222, 223
Glenmore Dam	viii, 18, 124, 174, 185, 190, 213
Glenmore Reservoir	112, 124, 138, 146, 163, 167, 171, 172, 178, 214
Graham, Bill	113, 118
grazing allotments	105
Grey Nuns	xiii, 183
greywater	203, 242
Griffith Woods	176, 177, 216
grizzly bear	13, 22, 23, 24, 44, 84, 90, 99, 218, 224
groundwater	xi, 16, 78, 107, 121, 125, 126, 127, 128, 130, 141, 144, 154, 167, 178, 198, 201, 203, 232, 236
Groundwater	236

H

hanging valleys	9
harlequin duck	100, 218
Harvey Barracks	166, 167, 215, 216

Index

Harvey, Brigadier Frederick	166
headward erosion	242
Heritage Park	113, 172, 214
Highland Stock Farms	152, 167
Highway 66	31, 52, 55, 67, 76, 77, 78, 98, 100, 104, 108, 109, 215
Holy Cross Hospital	181, 184, 210, 215
horse	32, 35, 38, 66, 67, 73, 83, 106, 108, 118, 119, 121, 122, 123, 133, 134, 136, 139, 142, 143, 144, 145, 146, 147, 148, 149, 151, 154, 160, 163, 178, 184, 185, 186, 187, 191, 196, 198, 209, 212, 218, 237
Howard, Ranger Terrence (Ted)	106, 122
Hudson's Bay Company (HBC)	135, 179
Husky Energy	90, 96, 97, 202
hydrologic cycle	16, 196

I

ice age	8, 242
I.G. Baker Company	179, 180
Imperial Oil	90, 212
Ings, Dr. George	75, 76, 210
Ings Mine	76, 79, 110
inlier	242
invertebrate	242

J

Jumpingpound Ridge	52, 54, 57

K

Kamp Kiwanis	116, 214
Kananaskis Country	13, 14, 22, 52, 59, 75, 93, 98, 108, 110, 114, 115, 117, 127, 155, 215, 222, 223, 224, 233
Kananaskis Country Campgrounds	108
Kananaskis Improvement District	xii, 108, 201

Kanouse, Fred	136, 179, 210
Kerr, Illingworth Holey	70
King's Own Calgary Regiment	158, 168
King, Thomas	145
Korean War	166
krummholz	21
kruppelholz	21, 55

L

Lacombe, Father Albert	181
Lake Rae	27
larch	21, 42, 43, 196, 217
Laurentide Ice Sheet	8
Leduc, Father Hippolyte	182
Leopold, Aldo	95, 109, 110, 224
Little Elbow Campground	33, 52, 57, 63
Little Elbow River	18, 29, 32, 35, 36, 46
Little Ice Age	11, 135, 242
Livingston, Sam	113, 136, 174, 179, 181, 210
Lord Strathcona's Horse (Royal Canadians)	158, 162, 164, 165, 166
Lougheed, Peter	13, 93
Lowes, Fred C.	185

M

MacEwan, Grant	xii, 174, 203, 224
Macleod, Lt.-Col. James	180
Macleod Trail	184
macroinvertebrate	38, 40, 47, 96, 242
Marquis of Lorne	33, 70
Matthews, Charlie	166
Matthews, Lt. Don	167
Matthews, Richard	166, 167
May, Captain Ernest	162

McCall, Fred	163
McLean Creek	91, 92, 93, 94, 96, 97, 98, 104, 124, 205
McLean Creek Off-Highway Vehicle (OHV) area	92
McLean Creek Public Land Use Zone	91
McLean, Jack	93, 116
McLean Pond	39, 93, 94, 215
meander cutoff	62
meandering stream	243
military	138, 157, 159, 160, 162, 163, 166, 167, 168, 169, 199, 213
Military Museums	169
Mission district	178, 181, 182, 183, 184, 185, 187, 200, 210, 211
Mitchell, Shorty	123
Moose Mountain	57, 60, 71, 72, 74, 75, 76, 77, 78, 80, 82, 83, 84, 89, 90, 98, 108, 109, 112, 115, 139, 210, 212, 214, 215, 235
Moosepacker's Trail	83
moraine	9, 29, 243, 244
mountain pine beetle	87, 88, 114, 219
Mount Burns	46
Mount Cornwall	31, 33, 35, 40
Mount Glasgow	31, 33, 34, 39, 46, 71
Mount Head	7
Mount McDougall	118, 214
Mount Rae	4, 5, 8, 27, 43, 44
Mount Remus	37
Mount Romulus	32, 36, 37, 39, 41
Mount Romulus campground	36, 44
Mount Rose	46, 58
Mount Royal College	19, 166, 215, 216
Mount Royal University	146, 216
mule deer	xiii, 99, 104, 219

N

Nihahi Ridge	33, 34, 46, 57
North-West Mounted Police	136, 137, 157, 158, 179, 180, 181, 182, 210
Northwest Rebellion	142, 147
Notre-Dame-de-la-Paix Mission	181

O

Oblate missionaries	181, 210
OHV trails	94, 96, 98
Opal Range	5, 7
Our Lady of Peace Mission	112, 137, 181
Outlaw Peak	31, 35, 42, 43, 44
oxbow lake	62

P

Paddy's Flat	75, 77, 102, 110
paleoclimate	10, 243
Palliser, Captain John	5, 14, 136
patterned ground	59, 80
Pattison, Private John	165
Pearce, Fred	203, 204
Pearce, William	148
Peigan	135, 136, 137, 210
periglacial	59, 243
pika	10, 12, 13, 54, 80, 196, 219
Piper, Norma	19, 22
Piper Pass	22, 23, 24
Pirmez Creek	116, 146, 147, 149, 150, 213
Pirmez Irrigation Society	210
Pirmez, Raoul	148
plunge pool	69, 244

Pocaterra, George 18
Potter, Mike 82
Powderface Ridge 52, 55, 66
Powderface, Tom 55, 106
Prairie Creek 73, 96
Prairie Mountain 57, 71, 72
Princess Patricia's Canadian Light Infantry 165, 166, 168
Provincial Recreation Areas 110, 222

Q

Quirk Creek 65, 90, 94, 215
Quirk Ridge 65, 67

R

Rae, Dr. John 5
Rae Glacier 4, 8, 9, 12, 18, 19, 29, 235
Rainy Creek Loop 66
Rainy Summit 65, 66, 67
Ranchmen's Club 184
RCAF Station Lincoln Park 165, 166, 214
reach xi, 19, 62, 63, 65, 68, 70, 100, 102, 127, 134, 135, 244
recessional moraine 244
Redwood Meadows 123, 140, 215
riparian zone 36, 37, 38
river profile 18, 19, 47, 77, 125
Riverwalk 191
Robinson, Joseph 158
Robinson, Richard George (RG) 146
Rock art (geoglyphs) 160
rock glacier 59, 60
Rocky View County xii, 112, 127, 128, 151, 152, 201, 229
Rouleau, Charles B. 182
Rouleauville 182, 183, 184, 185, 200, 210, 211

Round Hall	116
Roxboro	178, 182, 185, 211
Russell, Charles	190

S

Saddledome	190
saddles (cols)	9
Sandford, Robert	205
Sarcee City	159
Sarcee Gravel Products	139
Sarcee Training Camp	159, 160, 162, 166, 168, 211
Scollen, Father Constantine	137, 181
scree	34, 40, 58, 78, 80, 81, 244, 245
sensitive species	244
Service Flying Training School (SFTS)	165, 214
Sharpe, Sydney	224
Sheep Lakes	27
Sheep River	27, 60
Shell Oil	78, 90, 96, 202, 215
Signal Hill	140, 142, 162, 211, 215
Silvester Creek	95
Simpson, Sir George	179
Snowbirds Seniors Centre	130
South Fork	36, 37, 41, 42
Spray Lake Sawmills	88, 96, 97, 114, 202
Springbank	113, 116, 120, 124, 136, 143, 144, 146, 148, 149, 150, 151, 158, 162, 165, 167, 187, 188, 210, 212, 213, 214, 215
Springbank Irrigation Company	150
Spruce Vale	142, 147, 150
Stampede Park	178, 187, 190
Starlight, Bruce	189
Station Flats	76, 89, 110

Index

stone polygon	59
Stoney	14, 33, 55, 71, 106, 125, 135, 136, 137
Stoney/Nakoda	14, 135
stormwater	138, 172, 176, 177, 201, 202
stotting	104
Strange, Thomas	70
subalpine	5, 13, 14, 16, 17, 20, 21, 23, 24, 27, 29, 37, 38, 40, 41, 42, 44, 55, 58, 73, 80, 99, 104, 139, 144, 196, 220
subalpine zone	245
Sun Dance	139
Sustain Kananaskis	115
Suzuki, David	203

T

talus	17, 40, 43, 244, 245
Talus Creek	33
Talus Lake	39, 40
tarn	9, 29, 30, 40
terrace	78, 143
Thompson, David	136, 235
Threepoint Mountain	46
Tombstone Campground	29, 43, 44
Tombstone Lakes	24, 29, 41, 44
Tombstone Mountain	23, 27, 29, 40, 42, 43
Tombstone Pass	29, 40, 42
Trading Post	121, 124, 125, 126, 212
trout	17, 29, 39, 40, 71, 74, 93, 94, 113, 217, 218, 219
Trout Unlimited Canada	94, 96, 97, 225
Tsuu T'ina	xii, 113, 123, 127, 135, 136, 137, 138, 139, 147, 148, 159, 166, 167, 168, 174, 176, 187, 189, 202, 209, 210, 213, 214, 215, 216
Twin Bridges	116, 143, 162, 166
Two Pines	120, 213

U

unconfined aquifer	245
University of Alberta	54, 166, 213
University of Calgary	27, 34, 47, 223, 224, 236
Upper Elbow General Store	125
U-shaped valleys	9

V

Victoria Park	177, 186, 211
vulnerable species	245

W

Wake Siah Lodge	118, 119, 123
wastewater	vii, 128, 172, 177, 178, 202, 213, 214, 215, 242, 245
Water for Life	92, 93, 216, 221, 222
Watson, John "Gravity"	173
Weadick, Guy	187, 189
Weaselhead	138, 139, 171, 174, 176, 185, 202
West Bragg Creek	84, 89, 110, 111, 112, 114, 115, 116, 120, 223
wetlands	177
Wheeler, A.O.	112, 148, 210
White, Harry and Ida May	118
Whitney, Roy	189
Whymper, Edward	33
wild horses	66
wildlife management	95
winter range	67, 104
Wolf's Flat Ordnance Disposal	139
World War I	7, 33, 76, 106, 158, 159, 162, 163
World War II	164

X

XC Ranch   148

Y

Young, James, William and Thomas   149
Young, Percy   164, 165
Young, Rex   164
Young, Walter   164, 165
youth hostel   118, 213

CPSIA information can be obtained
at www.ICGtesting.com
Printed in the USA
LVOW05s2000261015
459849LV00017B/87/P